WJEC Level 1/2 Vocational Award
ENGINEERING
(TECHNICAL AWARD)

REVISED EDITION

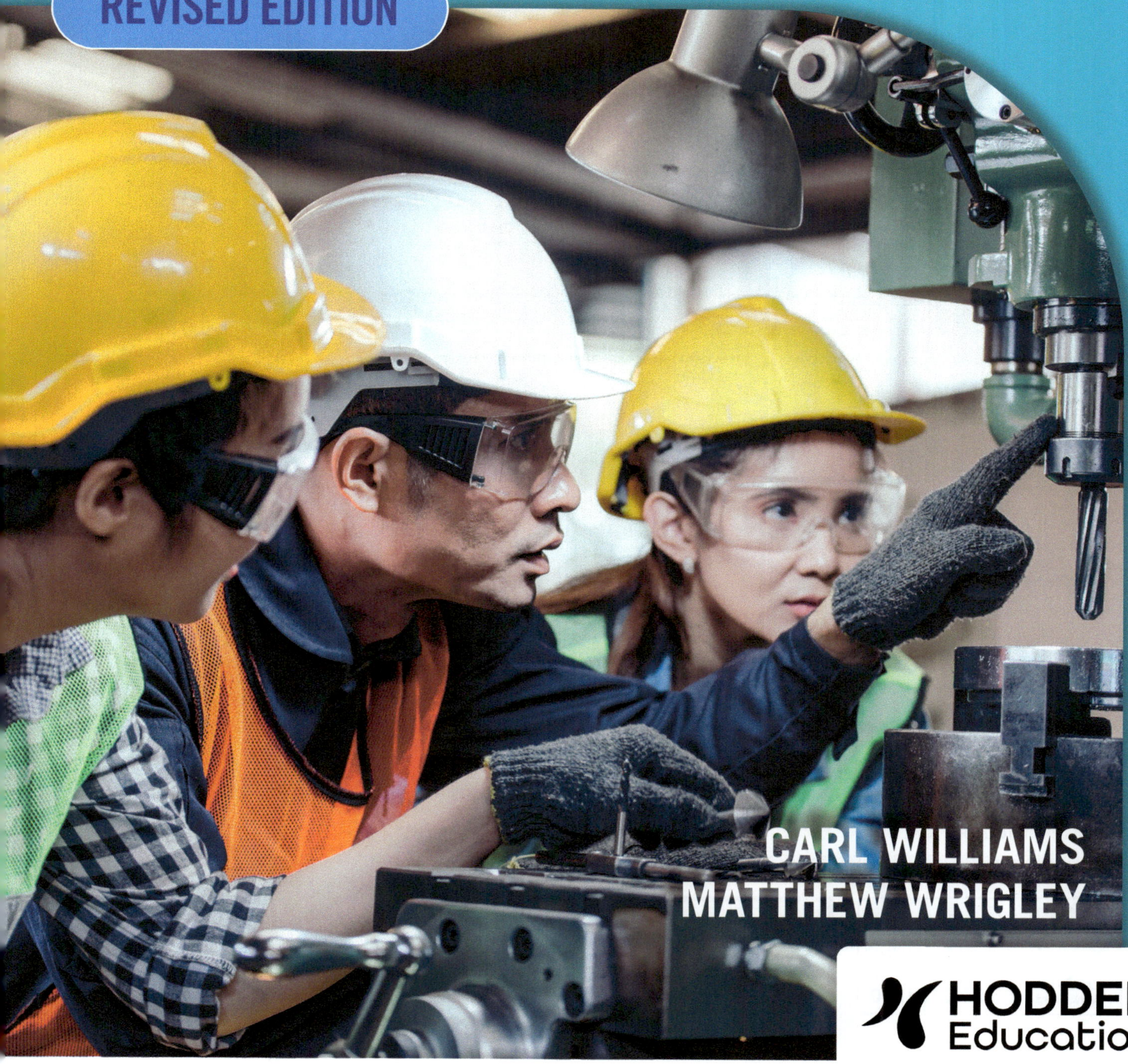

CARL WILLIAMS
MATTHEW WRIGLEY

HODDER Education

Although every effort has been made to ensure that website addresses are correct at time of going to press, Hodder Education cannot be held responsible for the content of any website mentioned in this book. It is sometimes possible to find a relocated web page by typing in the address of the home page for a website in the URL window of your browser.

Hachette UK's policy is to use papers that are natural, renewable and recyclable products and made from wood grown in well-managed forests and other controlled sources. The logging and manufacturing processes are expected to conform to the environmental regulations of the country of origin.

To order, please visit www.hoddereducation.com or contact Customer Service at education@hachette.co.uk / +44 (0)1235 827827.

ISBN: 978 1 3983 7951 0

© Carl Williams, Matthew Wrigley 2024
First published in 2019
This edition published in 2024 by
Hodder Education,
An Hachette UK Company
Carmelite House
50 Victoria Embankment
London EC4Y 0DZ
www.hoddereducation.com

Impression number   10 9 8 7 6 5 4 3 2

Year   2028 2027 2026 2025 2024

All rights reserved. Apart from any use permitted under UK copyright law, no part of this publication may be reproduced or transmitted in any form or by any means, electronic or mechanical, including photocopying and recording, or held within any information storage and retrieval system, without permission in writing from the publisher or under licence from the Copyright Licensing Agency Limited. Further details of such licences (for reprographic reproduction) may be obtained from the Copyright Licensing Agency Limited, www.cla.co.uk

Cover photo © winnievinzence – stock.adobe.com

Typeset by DC Graphic Design Limited, Hextable, Kent

Printed by Ashford Colour Press Ltd

A catalogue record for this title is available from the British Library.

Contents

Introduction	iv

1 Understanding engineering drawings — 1
- Introduction — 2
- Drawing standards — 2
- Organisation and equipment — 3
- Isometric drawing — 4
- Cutaway drawings — 12
- Exploded views — 12
- Orthographic projections — 13

2 Planning manufacturing — 26
- Stock-forms — 26
- Planning and sequencing — 30

3 Using engineering tools and equipment — 40
- Introduction — 40
- Engineering equipment selection — 41
- Engineering machines — 53
- Woodworking tools and machines — 60
- CAD/CAM — 62
- Health and safety — 66

4 Implementing engineering processes — 74
- Introduction — 74
- Marking-out — 74
- Joining materials — 76
- Shaping materials — 81
- Material finishes — 85
- Evaluate the quality of engineered products — 88

5 Guide to coursework submission: Unit 1 — 91
- Unit 1: submission guide — 92

6 Designing engineering products — 100
- Primary features of the given engineered product — 101
- Design briefs — 105
- Identifying features of other engineered products — 108
- Producing a design specification — 115
- Solving applied engineering problems — 117
- Displaying data — 137
- Evaluation — 138
- Generating a range of engineering solutions — 140
- Developing ideas through to a conclusion — 143

7 Understanding the effects of engineering achievements — 146
- Describing engineering developments — 146
- Explaining how environmental issues affect engineering applications — 156
- Existing and future engineering materials and processes — 160

8 Understanding properties of engineering materials — 165
- Understanding materials, their properties and their selection for specific purposes — 165
- Describe properties required of materials for engineering products — 180
- Explaining how materials are tested for properties — 183

9 Guide to coursework submission: Unit 2 — 187
- Unit 2: submission guide — 188

Glossary of key terms	195
Index	199

Introduction

This book has been written and designed to give you all the relevant information you will need to successfully complete the WJEC Level 1/2 Vocational Award in Engineering (Technical Award). It has been mapped to the course specification and ordered in such a way as to allow you to learn relevant skills and theory and then apply them to the assessed units.

This book will introduce you to many basic engineering skills and principles, and will provide you with a good understanding of the subject area. You will learn how to communicate effectively as an engineer, and learn practical skills and theoretical knowledge to allow you to successfully complete the required units.

What you will find in this book

While using this book you will come across several features that will benefit your knowledge and understanding of the subject area, including key terms, tip boxes and task boxes.

> **Key Term**
> A key word or phrase associated with the subject area and a specific technical term used to demonstrate your knowledge of the vocabulary used by engineers.

> **Top tip**
> Some quick-fire advice to help you complete or understand the current task.

> **Task**
> A small task or mini-project designed to embed the knowledge you have just learned. Also allows you to 'have a go' at applying the theory.

The route to success

To learn all the relevant skills to a standard where the top performance bands and higher grades become available to you, ideally, you need access to some specialist equipment:

CAD: computer-aided design is not needed to complete the course but does produce quality outcomes. CAD is used extensively in the engineering industries and is a skill that needs to be learned at some stage when training to be an engineer. There are many CAD packages that, as a student, you can download free for trial periods, as well as some full CAD packages that are on offer from some companies.

Workshop: you will need access to a workshop environment to apply your learned skills and knowledge. One of the fundamentals of engineering is the ability to manipulate materials by using processes and equipment that can only be found in a workshop, such as milling, turning and drilling. You can use hand tools or machinery to create prototypes, but understanding how to set up and use machinery safely and effectively is a fundamental skill of this course.

Course structure

The WJEC Level 1/2 Vocational Award in Engineering (Technical Award) is a vocational course that can be started in Year 9 and run over three years or started in Year 10 and run over two years. The students completing the course will be assessed with three units and graded from Level 2 Distinction* (Star) to Level 1 Pass. You can find out more about grading information by looking at the WJEC website and selecting the Engineering Qualification: https://www.wjec.co.uk

Below you will find examples of how the course will be graded and assessed.

Grading

The grading structure for the course is as follows:

Grade	Written as	Approximate GCSE grade equivalent
Level 2 Distinction* (Star)	D*	8/9
Level 2 Distinction	D	7/8
Level 2 Merit	M	5/6
Level 2 Pass	L2P	4
Level 1 Distinction*	L1D*	3/4
Level 1 Distinction	L1D	3
Level 1 Merit	L1M	2
Level 1 Pass	L1P	1/2

Units

The following units will be assessed:

Unit title (entry code)	Assessment	Content
Unit 1 Manufacturing engineering products (5799U1)	Internal (school/college)	The assignment brief will include a scenario and several tasks
Unit 2 Designing engineering products (5799U2)	Internal (school/college)	The assignment brief will include a scenario and several tasks
Unit 3 Solving engineering problems (5799U3)	External (WJEC moderators/examiners)	Written examination

How will you be assessed?

You will be assessed with three units:

Unit 1: internally assessed (guided learning hours: 48)

Unit 1 provides you with the opportunity to interpret different types of engineering information to plan how to manufacture engineering products. You will develop knowledge, understanding and skills in using a range of engineering tools and equipment to manufacture and test an end product. The assignment brief will include a scenario and several tasks, which will be assessed/marked by your tutor.

This unit is a controlled assessment, with your tutor supervising the work.

You will have 20 hours to produce the assessable work. It will be marked out of 80, and contributes 40% to your overall grade.

Unit 2: internally assessed (guided learning hours: 24)

Unit 2 allows you to explore how an engineered product is adapted and improved over time, and it offers the opportunity to apply your knowledge and understanding to adapt an existing component, element or part of the engineering outcome that you manufactured for Unit 1. This will be assessed/marked by your tutor.

This unit is a controlled assessment, with your tutor supervising the work. It will be marked out of 40, and contributes 20% to your overall grade.

You will have 10 hours to produce the assessable work.

Unit 3: externally assessed (guided learning hours: 48)

Unit 3 is an externally assessed examination that you will sit in a controlled environment such as a school or college, supervised by invigilators. The examination will last 90 minutes and is worth 80 marks (40% of your total grade). It will cover all aspects of the engineering specification, including knowledge and theory.

When should you sit the units?

Submitting work for all three units should be dictated by your school or college, and your tutor will give you notice of when work should begin and when the deadlines are imminent.

To successfully complete this course you will need workshop practice. You will need to complete several workshop-based projects that will take you through procedures on how to use tools and equipment safely and accurately.

Your progress

Below is a handy checklist for you to go through every time you complete a chapter of this book. Copy the checklist into your notebook and tick the boxes when they are completed, and then put a tick in either the 'Poor', 'Okay' or 'Good' box to check your understanding of the chapter. When looking back over this checklist while revising, you will quickly be able to see which areas you are strong in and what areas you will need to revise further.

Chapter title	Tick if covered	Understanding		
		Poor	Okay	Good
1 Understanding engineering drawings				
2 Planning manufacturing				
3 Using engineering tools and equipment				
4 Implementing engineering processes				
5 Guide to coursework submission: Unit 1				
6 Designing engineering products				
7 Understanding the effects of engineering achievements				
8 Understanding properties of engineering materials				
9 Guide to coursework submission: Unit 2				

Understanding engineering drawings

1

In this chapter you are going to:
- learn about standardisation
- discover how to use standards when creating technical drawings
- learn to use the standardised conventions used in engineering drawings
- learn how to create:
  - isometric views
  - exploded views
  - third-angle orthographic projections
  - sectional/offset sectional views
- learn how to generate a range of engineering solutions and develop ideas through to a conclusion.

This chapter will cover the following areas of the WJEC specification:

Unit 1 Manufacturing engineering products: 1.1 Understanding engineering drawings		
• 1.1.1 Interpreting engineering drawings	• 1.1.2 Interpreting engineering information	• 1.1.3 Presenting engineering information

Unit 2 Designing engineering products: 2.2 Proposing design solutions		
• 2.2.1 Generating a range of engineered solutions	• 2.2.2 Developing ideas through to a conclusion	• 2.2.3 Communicating design ideas

Unit 2 Designing engineering products: 2.3 Communicating an engineered design solution	
• 2.3.1 Producing an engineering specification	• 2.3.2 Drawing an engineering design solution that adheres to recognised standards

Unit 2 Designing engineering products: 2.4 Solving applied engineering problems		
• 2.4.1 Using mathematical techniques for solving applied engineering problems	• 2.4.2 Justifying suitable materials for use in the final engineered solution	• 2.4.3 Justifying suitable processes for manufacturing the final engineered solution

Unit 3 Solving engineering problems: 3.4 Solving engineering problems	
• 3.4.1 Using mathematical techniques for solving engineering problems	• 3.4.2 Understanding and producing engineering drawings

Introduction

▲ *The drawings produced by engineers are used to manufacture products from PlayStations to aircraft to skyscrapers.*

Engineers constantly create and use drawings as part of their day-to-day job. Drawings allow engineers to physically 'see' the shape of a product, look at how something could be assembled, recognise the different views of a product, as well as identify important points such as dimensions, materials and hidden details.

The drawings produced by engineers are used to manufacture products from PlayStations to aircraft to skyscrapers, all of which need to be highly accurate. Any mistake in the drawings would be transferred to the actual product during the manufacturing stage and huge amounts of time and money could be lost. This is why all modern technical drawings need to be **standardised**.

By learning how to create engineering drawings, you will gain the ability to look at another engineer's drawings and interpret them. This new ability will include identifying what the dimensions in a drawing relate to, understanding what all the symbols mean, understanding what the different types of lines mean and being able to understand the writing on each drawing that is used to explain further details.

This ability to interpret technical information from engineering drawings is a skill that you will use in Unit 1 when you are asked to interpret engineering drawings supplied by the exam board.

Drawing standards

There are a number of organisations across the globe responsible for standardising the many products and services that we use, such as plug sockets, paper and technical specifications for industrial processes. UK-based engineers have to conform to the standards from two organisations:
- the **International Organization for Standardization (ISO)**
- the **British Standards Institution (BSI)**.

Both ISO and BSI have developed a recognised format to standardise technical drawings. If an engineer in the UK produces a set of technical drawings for a product to be manufactured in a factory in, say, China, then a set of standardised drawings would be needed. Using a

▲ *The ISO logo*

standardised drawing would enable the Chinese manufacturers to understand the technical information clearly and accurately. Drawings created with ISO and BSI standards are recognised throughout the world. ISO and BSI often work together to create standards. BSI often create the first issue of standards, usually a Publicly Available Specification (PAS) or British Standard (BS), which can in time be developed into an international (ISO) standard. When a product is assessed, tested and certified to meet the standards and certification requirements outlined by the BSI, products can achieve the **BSI kitemark**. This can then be displayed on the product and product packaging to indicate the superior safety, quality, digital security or sustainability of the product for both compliance and marketing purposes.

The standardisation numbers for technical drawing are:
- ISO 128
- BSI 8888:2017

▲ *The BSI kitemark*

Organisation and equipment

To learn the skills needed to produce technical drawings, it would be helpful to have access to some equipment. The list below gives some suggestions.

Drawing
- 1 × set of graphics pencils (3H to 3B) for drawing and sketching
- 1 or 2 black fineliner pens for picking out lines in a drawing or sketch
- 1 × steel rule (or a standard ruler)
- 1 × 180° protractor
- 1 × 45° set square
- 1 × 30° set square
- 1 × compass
- 1 × eraser
- 1 × A3 sketchbook (lower quality is OK)

Organisation
- 1 × A4 ring binder (with or without poly pockets)
- 1 × A3 plastic portfolio

> **Key Term**
>
> **The BSI Kitemark** For more than 120 years, the BSI Kitemark™ has been recognised as a symbol of outstanding quality, safety and trust across a wide range of products and services. Kitemark certification confirms that a product or service's claim has been independently and repeatedly tested by experts, meaning that you can have trust and confidence in products and services that are BSI Kitemark certified.

The basics

As well as understanding how your equipment works, it is important to follow some basic rules when drawing in engineering. Here are a few tips to get you started.

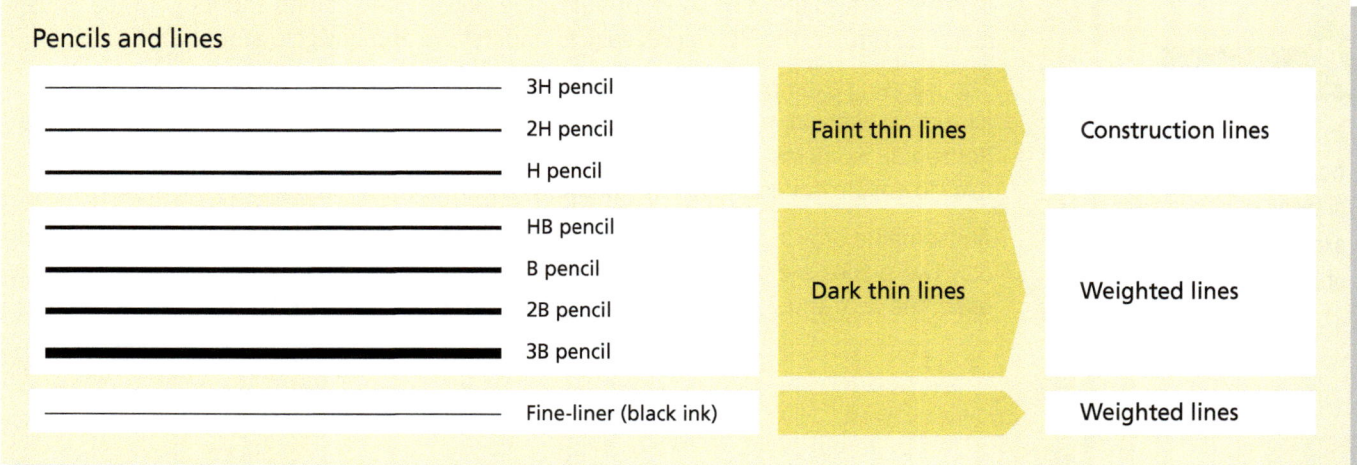

▲ *Different pencils are used to create the various types of lines used in engineering drawings.*

Key Terms

Construction lines Faint, thin lines that are easy to rub out and that can be used as a guide. They are drawn with a hard (H) pencil.

Weighted lines These lines define the object you are drawing, making it easier to see which lines to keep and which to erase. They are drawn with a soft (B) pencil or a fine-liner.

Top tip

When constructing a drawing, remember the phrase: 'light-is-right'.

Task 1.1

On an A3 sheet, label the lines your different pencils make. Then construct some simple shapes and pick them out using weighted lines.

Key Term

Axis The direction of travel from a fixed point. In 3D drawing there are three axes: X, Y and Z. (The plural of axis is axes.)

When creating a drawing you will first need to construct the overall shape with **construction lines**. These are faint, thin lines that are easy to rub out.

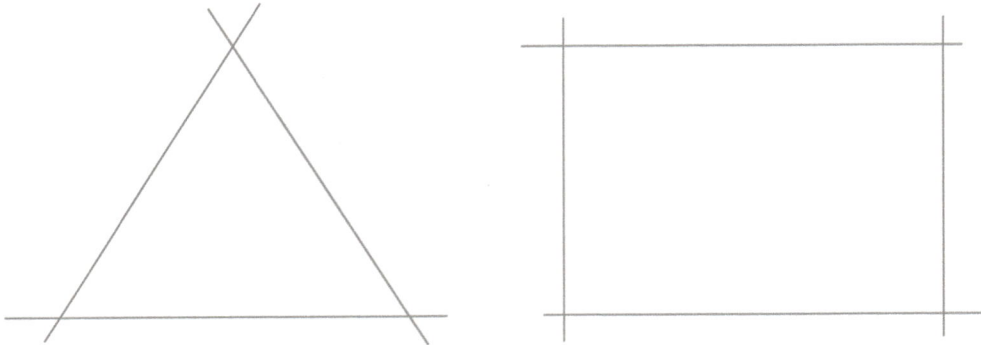

▲ *A triangle (left) and rectangle (right) constructed using construction lines.*

When 'picking out' the shape of your drawing, you will use **weighted lines**. The weighted lines define the object you are drawing and make it easier to see which lines to keep and which lines to erase.

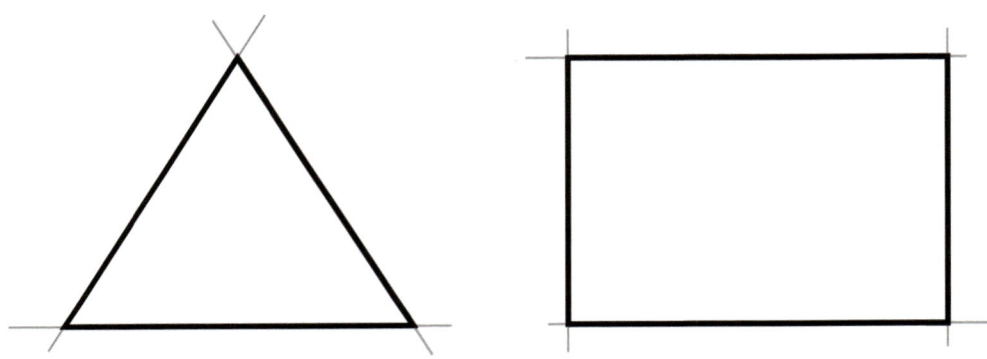

▲ *A triangle (left) and rectangle (right) picked out using weighted lines.*

Isometric drawing

Isometric drawing is a standardised (ISO, BSI) way of presenting designs and drawings in 3D. It is also a 'formal' way of presenting images in 3D and is used in many different ways to communicate information such as technical details and assembly instructions. Isometric is also the view used for many 'top-down' video games and for CAD (computer-aided design), as the 3D view is easy to understand and navigate around.

The following example has been drawn with a 30° set square. In isometric projection all vertical lines on an object remain vertical while all other lines are drawn at 30° to the horizontal. Isometric drawings are usually produced with drawing equipment or using CAD to ensure accuracy. When starting, you can use isometric grid paper to help.

An isometric drawing is constructed using three planes. In two-dimensional (2D) drawing only two planes are used: the X and Y **axes**. However, in 3D drawing, three planes are used: the X, Y and Z axes, with Z representing the third dimension.

1 Understanding engineering drawings

▲ Isometric grid paper can be used to produce standardised design drawings.

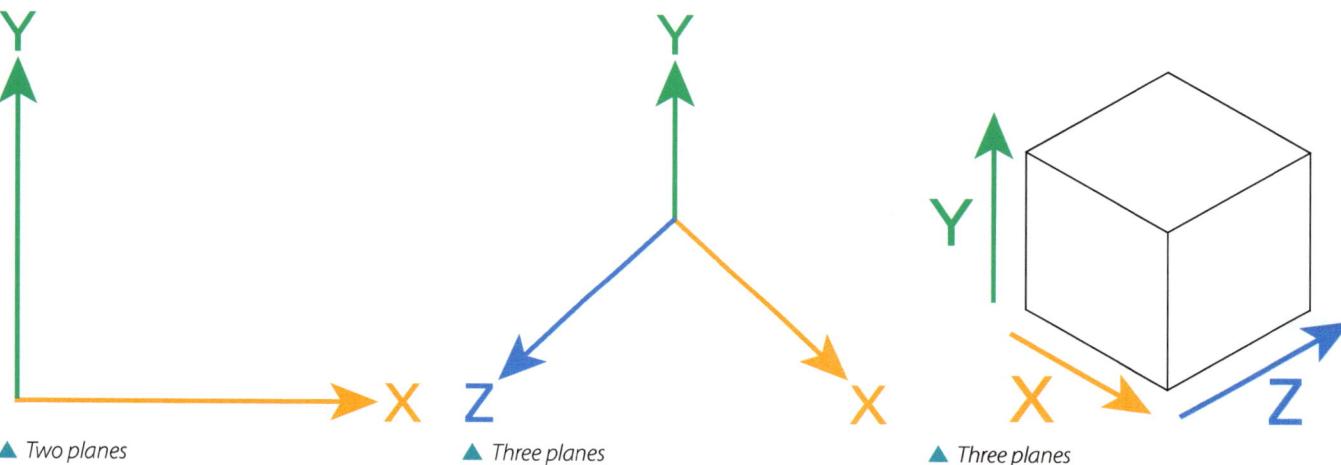

▲ Two planes   ▲ Three planes   ▲ Three planes

Constructing isometric drawings

In this section, we will construct our first isometric cuboid. You will need: an H pencil for construction lines, a B pencil for weighted lines, an A3 piece of paper (landscape) and a 30° set square. Constructing a cuboid is also known as constructing a **crate**.

Key Terms

Crate The name for the 3D 'box' you start your isometric drawings with.

Baseline The horizontal line you use to 'level' your set square.

Task 1.2

On your A3 sheet of paper, follow the guide below to create your first isometric crate.

1 Using your H pencil, draw a horizontal line to form your **baseline**.
2 Using your 30° set square, draw a vertical line from the centre of your baseline.
3 Draw a 30° line as shown.

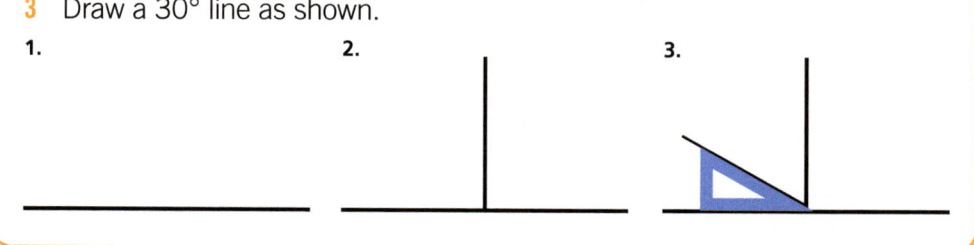

4. Reverse your set square and draw another 30° line intersecting with the first 30° line.
5. Raise your set square and draw another 30° line.
6. Reverse your set square and draw a further 30° line intersecting with the previous 30° line.

4.

5.
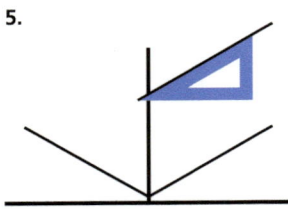
6.
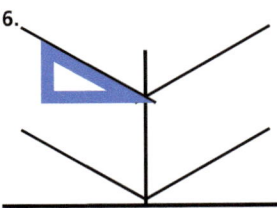

7. Rotate your set square and draw a vertical line.
8. Move your set square and draw another vertical line.
9. Rotate your set square and draw a further 30° line intersecting with the vertical line.

7.

8.
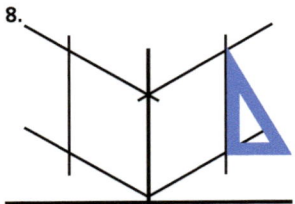
9.
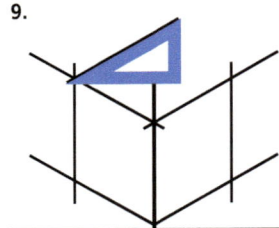

Top tip

All lines on the same plane should be parallel.

Top tip

Do not rub out your construction lines when the drawing is complete. They show your 'working out', just like in mathematics.

10. Reverse your set square and draw a 30° line intersecting with the previous vertical line.
11. Now you have a completed isometric cuboid drawn in construction lines.
12. Pick out your 3D shape with weighted lines.

10.
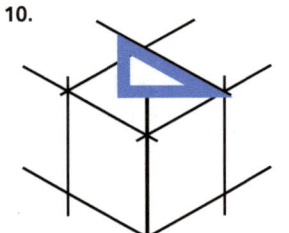
11.
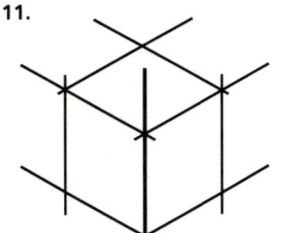
12.
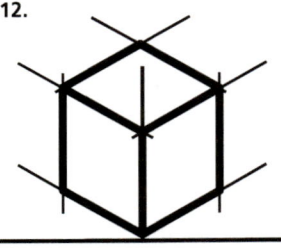

Constructing shapes in isometric

Now you know how to construct an isometric cuboid, or crate, you can use that space to create other shapes.

When artists create realistic images, they often build up the image layer upon layer by understanding how one layer affects the next (for example, understanding an animal's physiology to build the skeleton, muscles, skin and hair/fur). Engineers are more like sculptors; they start with a block of material and remove all the unwanted material to leave the shape needed. (You might imagine the crate as a block of ice or marble.) The following examples are some of the ways of removing material from a crate to produce the desired image.

Removing material

1. Construct a crate in isometric.

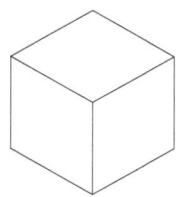

2. Trim your crate to the desired size using vertical and 30° lines.

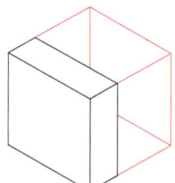

3. Identify the shape you need and remove the excess material.

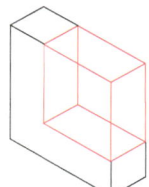

4. Pick out your shape with weighted lines and remove the construction lines.

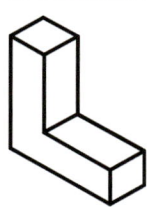

5. Finish with shading or rendering if needed.

Top tip

Trust your vision while you are constructing your isometric drawing. If it looks wrong, then it is wrong (so fix it).

Extruding shapes

You can also use isometric cuboids to 'extrude' profiles through one **plane** to create 3D objects. **Extrusions** are profiles that have been extended or stretched, much like prisms.

1. Choose a face on your isometric crate and draw the desired shape. Using 30° lines 'extrude' your shape onto the back/rear face of your isometric cube.

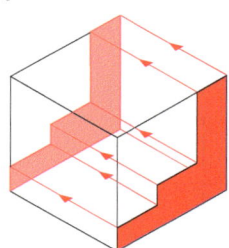

2. Join the two profiles using 30° lines to the detail points (corners, etc.).

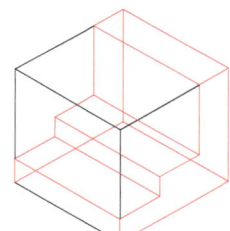

3. After picking out your shape in weighted lines, remove construction lines and finish as needed.

Task 1.3

In your sketchbooks, draw isometric drawings of the 3D shapes below.

Make sure you start this exercise by constructing isometric crates and working within the crate 'space'. Your drawings **do not** have to be dimensionally accurate.

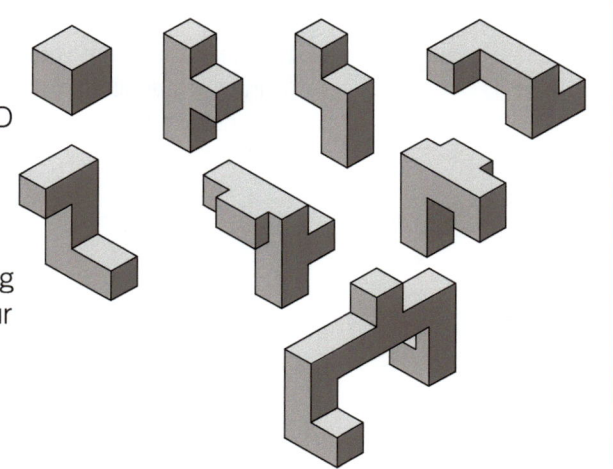

Key Terms

Planes The X, Y and Z axes (directions) in which you create.

Extrusions Profiles that have been extended or stretched.

Angles and curves in isometric

Angles

Not all the products we see and use are constructed from squares, flat edges and cuboids. The vast majority of items we see and use every day are made up from angles, curves and circles, as well as squares and cuboids. Engineers need to be able to communicate all shapes clearly and effectively, including shapes with angles and curves.

In this section we are going to look at how to construct angles in isometric.

Angles in one plane

1 Construct a crate in isometric.

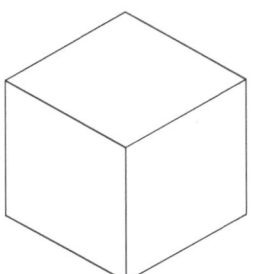

2 Choose a face, then draw an angled line from one corner to the opposite corner. Now extrude the angled line to the opposite face of the crate.

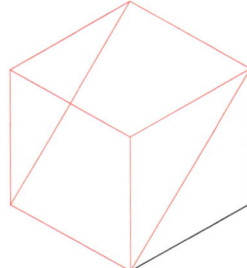

3 Pick out your angled shape using weighted lines and rub out the construction lines.

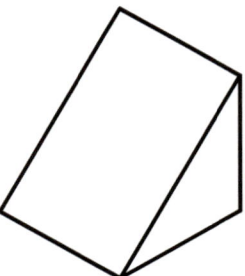

4 Finish your new angled shape with shading if needed.

Angles in two planes

1 Construct a crate in isometric as before. Choose a face and extrude an angled line through one plane.

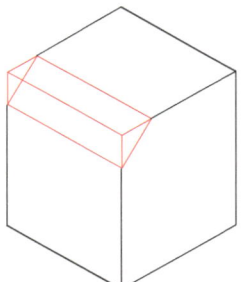

2 Repeat Step 1 but extrude your angle through a second plane.

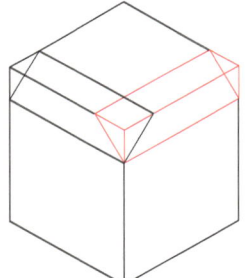

3 Repeat Steps 1 and 2 but on the base of your isometric crate.

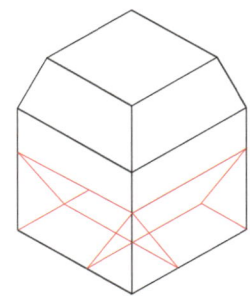

4 After picking out your shape in weighted lines, rub out the construction lines and finish with shading if needed.

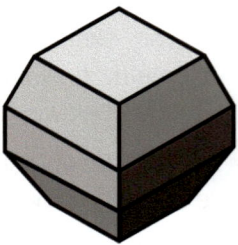

Task 1.4

On an A3 sheet, construct three crates and draw shapes that include angles. Try to use all three planes for your final drawing.

Circles and curves in isometric

When drawing circles and curves in 3D, you have to take into account the **perspective** of an object. Isometric uses something called an **axonometric perspective**. An axonometric perspective (isometric view) is a 'pictorial' representation of 3D objects and not a true view of how we view the world. True perspective shows lines disappearing into the distance and converging at one or more points (**vanishing points**), while isometric drawing shows lines running parallel (30° lines). Axonometric projections have the advantage of being clear to understand as they are used in recognised drawing formats (ISO, BSI) such as isometric. In addition, they are accurate and allow dimensions to be added.

Key Terms

Perspective Your point of view when looking at an object.

Axonometric perspective A pictorial representation of a 3D object that is not a true view of how you would view it.

Vanishing points Lines that disappear into the distance.

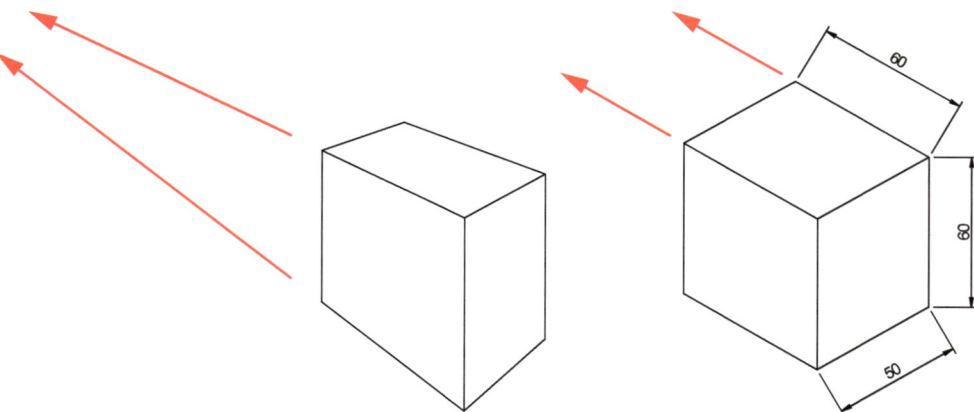

▲ Examples of perspective. The image on the left is drawn in true perspective, with parallel lines converging to a vanishing point; the right-hand image is an example of an axonometric drawing. The lines are parallel so you can add dimensions and give the viewer more technical information. Lines converge in perspective.

Circles in isometric drawing are known as **ellipses**. When looking at a cylinder (for example, a tin of beans or cola can), you know that the top and bottom of the cylinder are circles. However, what you see is in fact an ellipse.

Key Term

Ellipse A circle viewed in axonometric projection.

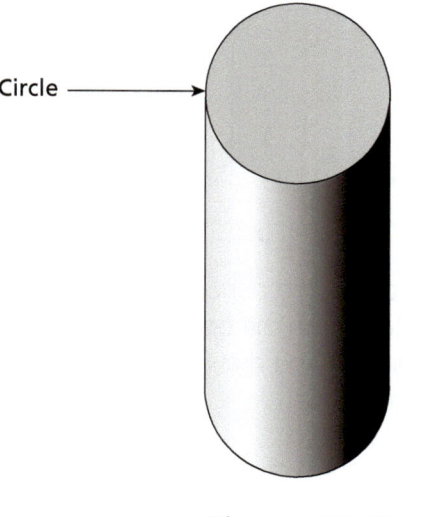

▲ Which of these two 3D drawings looks correct?

When discussing ellipses we often talk about the **major axis** and the **minor axis**. The diagram on the right illustrates this idea.

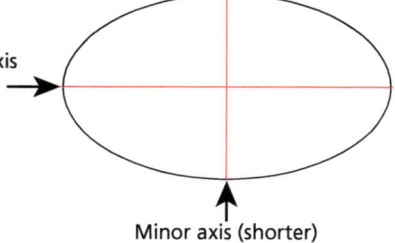

▲ Major and minor axes

Ellipses in isometric

There are several methods of drawing ellipses accurately, using basic drawing equipment such as rulers, set squares, compasses and pencils. These methods are detailed in the following table.

> **Key Term**
>
> **Trammel** A trammel of Archimedes (also known as an ellipsograph) is a device that can be used to draw ellipses. The trammel method can also be replicated by using a piece of paper and a major and minor axis of the ellipse to be drawn.

The freehand-sketch method	Used when sketching out ideas or practising your isometric drawing. Only a pencil is needed for this method.
The concentric circle method	Two concentric circles are used with the major and minor axes of the isometric 'diamond'. A compass, ruler and pencil are needed for this method.
The trammel method	Uses the major and minor axes of an isometric diamond as well as a strip of paper used as a *trammel*. A ruler, pencil and trammel are needed for this method.
The four-centre method	Uses the original circle with a series of centre points and drawn arcs to create an isometric ellipse. A ruler, compass and pencil are needed for this method.

Task 1.5

Using the four-centre method guide shown below, construct an ellipse on an A3 sheet of paper.

Four-centre method

1. Draw a circle with the required diameter (for the major axis of the ellipse) within a square.
2. Quarter the circle/square.
3. Draw a line from point **A** to point **B** and from point **C** to point **D**.
4. Repeat the process but from the opposite corners.

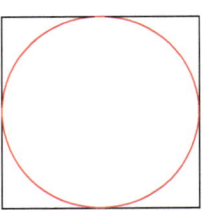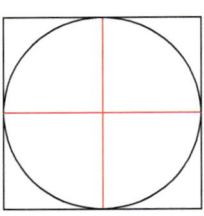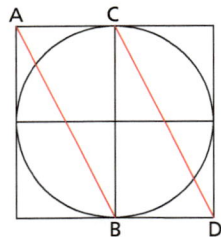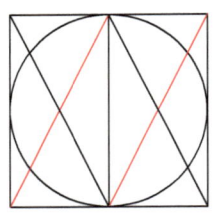

5. Using a compass, set the metal centre to point **E**, the pencil to point **F** and draw an arc between points **F** and **G**. Repeat on the opposite side.
6. Using a compass, set the metal centre to point **H** and draw an arc between points **I** and **J**. Repeat on the opposite side.
7. Pick out your ellipse with weighted lines and finish.

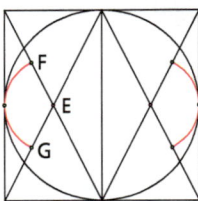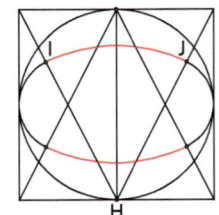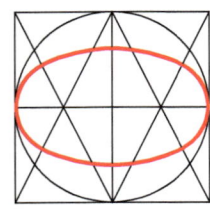

1 Understanding engineering drawings

When drawing ellipses in isometric, you can draw in one of the three different planes (**X**, **Y** and **Z**).

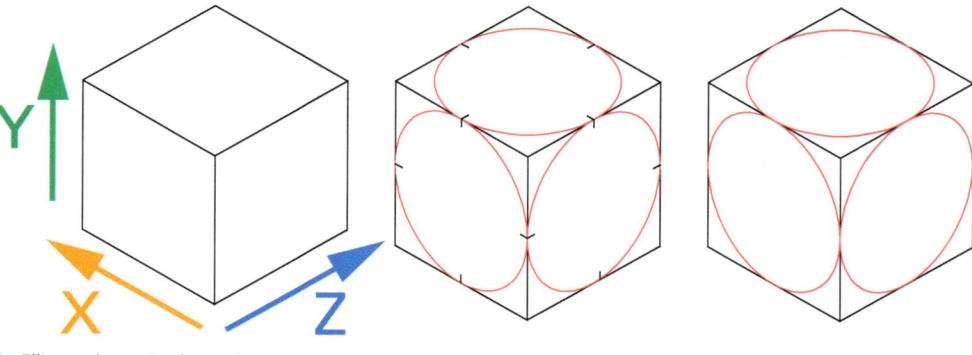

▲ Ellipses drawn in three planes

▲ A cylinder created by drawing two ellipses in one plane.

When sketching freehand in isometric to generate quick ideas, or even when discussing ideas with clients and customers, you can follow these simple rules:

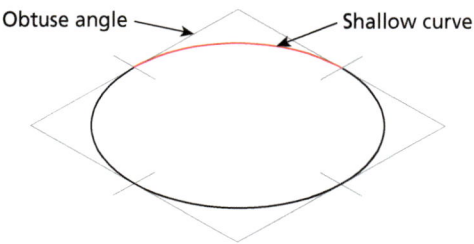

▲ Draw a shallow curve when you have an obtuse angle.

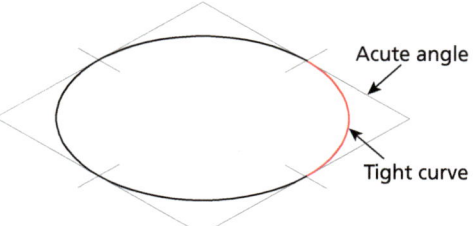

▲ Draw a tight curve when you have an acute angle.

Key Term

Fillets Corner curves.

Task 1.6

Drawing a trendy coffee table

Copy the drawings below into your notebook then, using the correct equipment (set square and ruler), complete the drawing exercise, demonstrating the skills you have learnt so far. You can draw the **fillets** freehand.

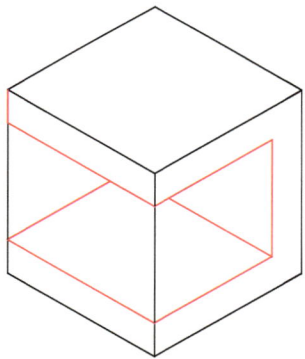 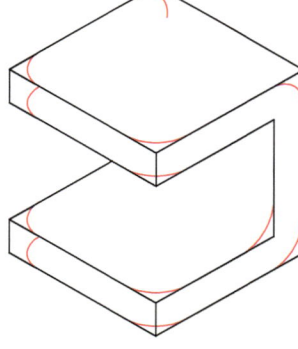 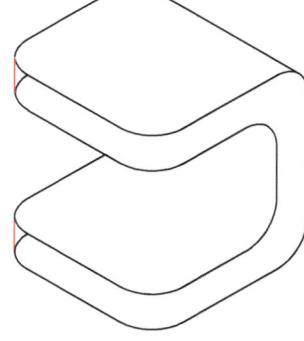 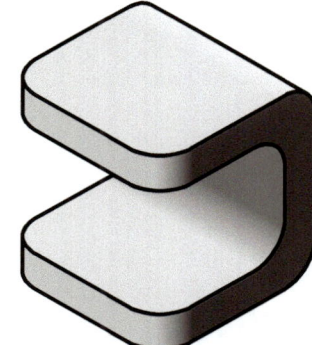

1 Draw an isometric crate and remove most of the middle section.

2 In freehand, draw sections of ellipses on the two different planes until you have created fillets on the corners and edges.

3 Connect the corner fillets and rub out the corner lines that have fillets.

4 Pick out the coffee table shape with weighted lines and shade.

11

Cutaway drawings

Cutaway drawings are designed to show the viewer important parts of the interior of an opaque object or product (an object that you cannot see into because it has a solid exterior/case). This is achieved by 'cutting away' parts of the exterior and leaving other parts of the exterior intact. By doing this, you can show many important features of products such as the internal layout of the seating on an aeroplane, the pistons moving in an engine or how the internal components of a drill fit into the casing. Cutaway drawings also show the different parts of a product and how one part can interact with another.

Constructing a cutaway drawing

When constructing cutaway drawings, you are trying to communicate information to the viewer that can sometimes be very complicated. By following a series of tried-and-tested guidelines, the finished drawing will not only be easier to construct but also easier to understand.

Guidelines
- Construct your cutaway drawing in isometric.
- When drawing your product, think of it as different parts, not a whole product.
- Only part of the exterior will be 'cut away'.
- All parts that have been 'cut' will be **hatched**.
- When two different parts meet, try to draw hatching in the opposite direction.

> **Key Term**
>
> **Hatching** A series of 45° parallel lines that are separated by an appropriate distance (e.g. 4 mm) to show where a solid object has been cut.

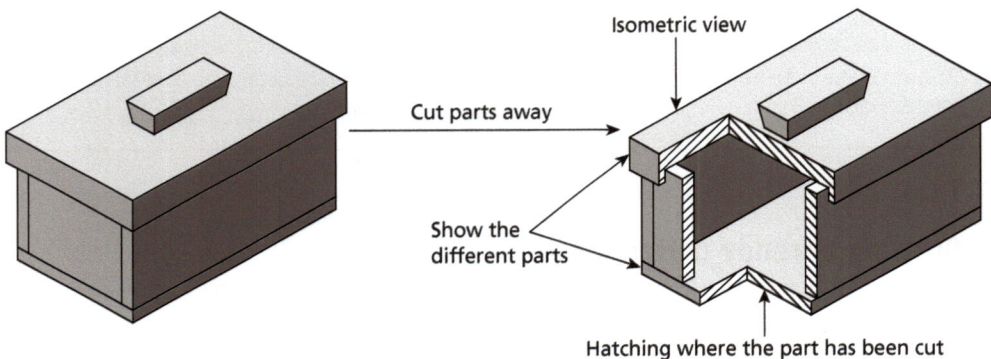

▲ An example of a cutaway drawing

> **Task 1.7**
>
> Find a simple object from your house (a jewellery box, pen, drinking bottle, etc.) and produce an isometric cutaway drawing.

Exploded views

Exploded drawings (exploded views) are created to show all the different parts of a product and how they are **assembled**. If you have ever used an instruction manual to assemble furniture or certain toys, you have probably seen exploded drawings. They are a great pictorial aid for showing how all the component parts of a product **interrelate**.

At first glance, exploded drawings look quite complicated. In fact, they are simple isometric drawings that use only the drawing skills demonstrated so far in this chapter.

> **Key Terms**
>
> **Assembled** Put together.
>
> **Interrelate** To communicate with other people or, when talking about component parts, to work together.

Constructing an exploded view

Imagine a product is exploding … then pause it in time. The result should show all the parts slightly separated. If you then slowly reverse the explosion, you can see how each of those parts go together. Below you will find some useful tips on how to create an exploded drawing:

- Always draw your exploded drawing in isometric.
- Think of it as different parts, not the whole product.
- Use the same projection lines for parts that are opposite each other.

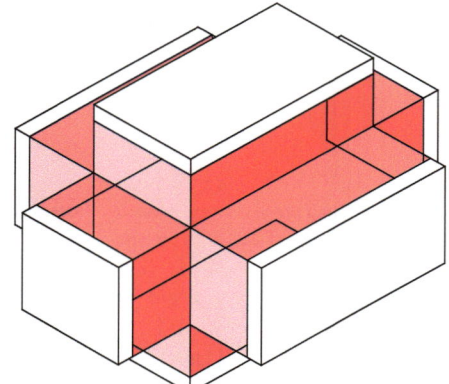

▲ This exploded drawing has been constructed using three isometric crates. The three crates all 'intersect' in the middle. All three axes are used when projecting or extruding the crates.

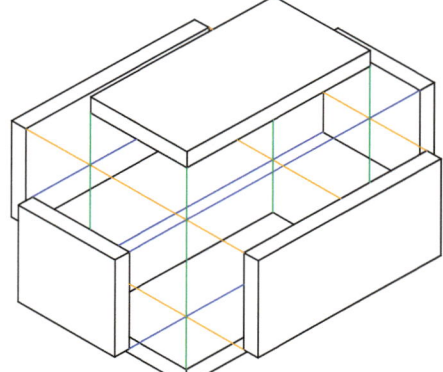

▲ This exploded drawing has been constructed using projection lines. See how one part has been drawn and then extruded using projection lines. The parts have been projected along the X, Y and Z planes.

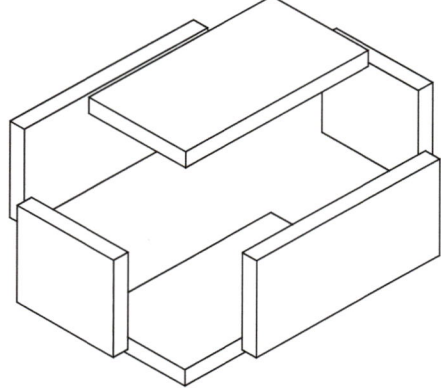

▲ When you erase the construction lines and use weighted lines to pick out your drawing, you will end up with a completed, easy-to-understand exploded drawing.

Orthographic projections

Orthographic projections are standardised drawings (ISO, BSI) that contain all the relevant technical information needed for the part or product to be made by a third party. Engineers regularly design parts and products that will be manufactured elsewhere, often in other countries. The drawings need to be extremely accurate with all the vital information communicated clearly and efficiently. By having standardised drawings, it means that anyone reading the drawing will be able to understand it, as it will conform to the relevant standards from ISO and BSI. Orthographic projections are often known as **technical drawings**, **working drawings** or **engineering drawings** and can contain lots of relevant **conventions** such as:

- different views
- dimensions
- scale
- materials
- hidden detail
- centre lines
- finishes
- section views
- the date the drawing was produced
- the name of the engineer or designer
- the 'angle' symbol
- a title
- a parts list
- detail of manufacturing processes.

Orthographic projections are constructed using different **views** of the part or product. This enables the viewer to see details that might sometimes be hidden. The views that are usually shown are:

- **front view**
- **side view**
- **plan view**
- sometimes an isometric or section view.

Key Terms

Orthographic projection In engineering, this is a means of representing different views of an object by projecting it onto a plane or surface.

Technical drawings The common term used for third-angle orthographic projections.

Conventions Technical terms.

Views Views are also known as elevations.

Plan view A view from above, also known as a bird's-eye view.

WJEC Level 1/2 Vocational Award Engineering (Technical Award)

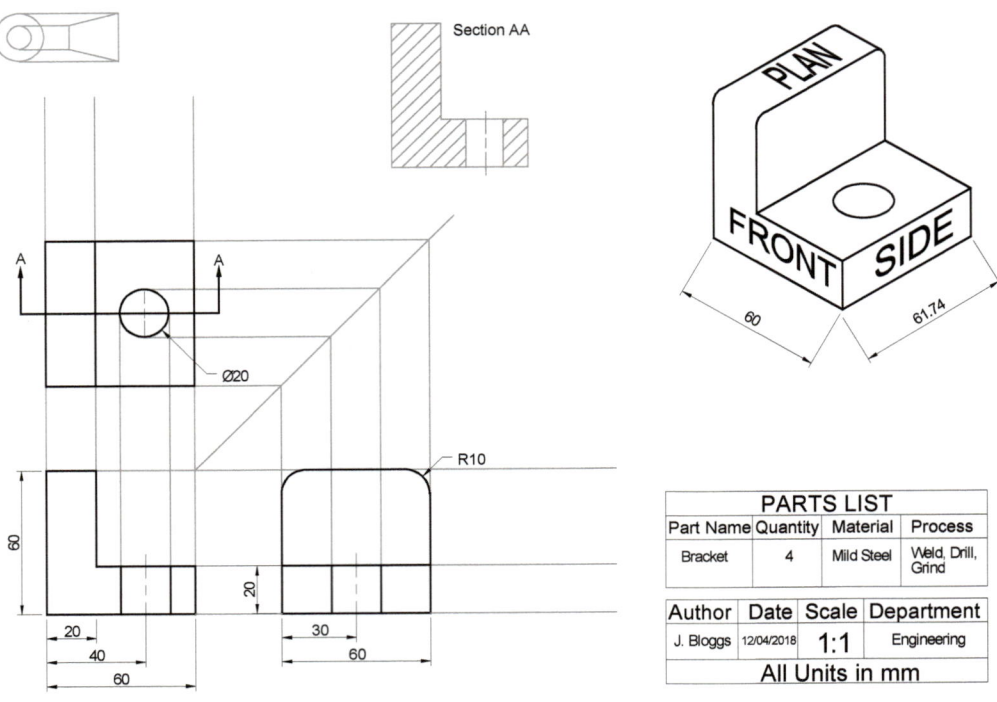

▲ Example of an orthographic projection

There are two different methods of drawing in orthographic projection: first angle and third angle. These methods differ in the views you choose to show.

Below are examples of a first- and third-angle orthographic projection that show how the different views of a 3D part or product are laid out.

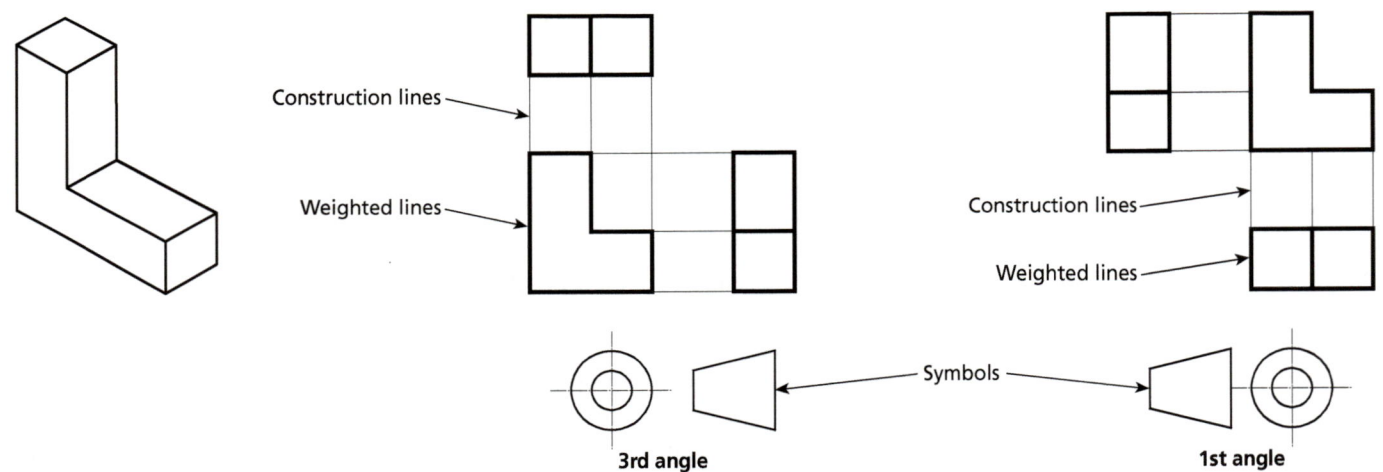

▲ This 3D shape (isometric) is being shown in the orthographic projections on the right.

Key Term

Representations Views.

As well as **representations** of the 3D shape, you can also see the 'symbols' used to show what type of drawing is being displayed (first or third angle). The symbol used is based on a 3D shape that looks like a solid lampshade or a cone with the top cut off.

1 Understanding engineering drawings

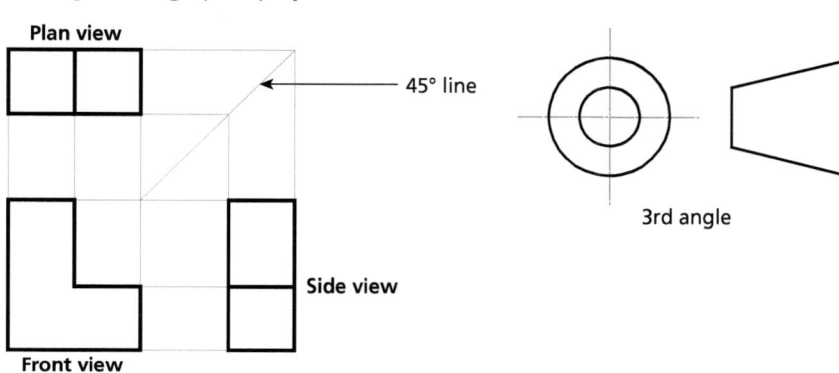

▲ 3D symbol shape

Third-angle orthographic projections

In this section, you will learn how to draw in third-angle orthographic projection.

The way you 'rotate' the 3D object to create the views in third-angle orthographic projection is very specific. You must turn the object 90° in the correct direction.

The correct position of each view is shown on the right. The end result must be very accurate and should be completed using the correct drawing equipment or CAD.

The diagram below shows how the 3D object is rotated to display each of the views for third-angle orthographic projection.

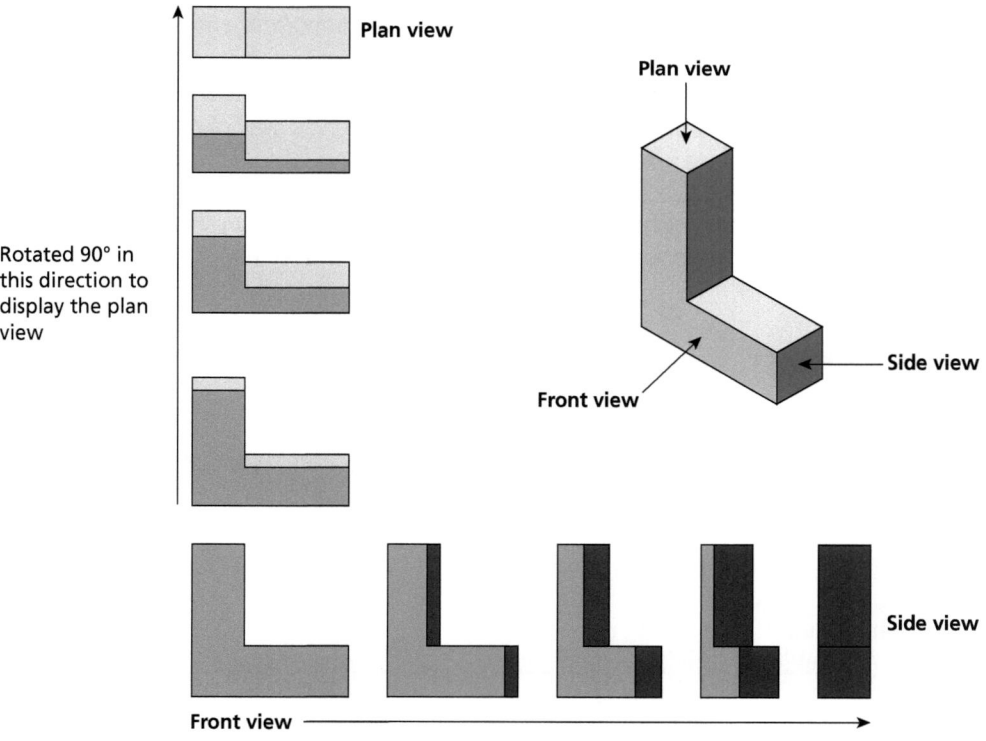

▲ The 3D object showing all the views that will be displayed.

Top tip

Spend some time getting used to *how* the object moves from the original (front) view, as seen in the drawings in this section. Practise drawing the different views of simple shapes.

Task 1.8

Draw or find a few simple 3D objects and construct third-angle orthographic projections of them on an A3 sheet of paper. The drawings do not have to be dimensionally accurate. Try to get the views in the correct positions (front, plan, side).

The following diagram shows how the 3D object is projected onto a 'wall' to display each view for third-angle orthographic projection.

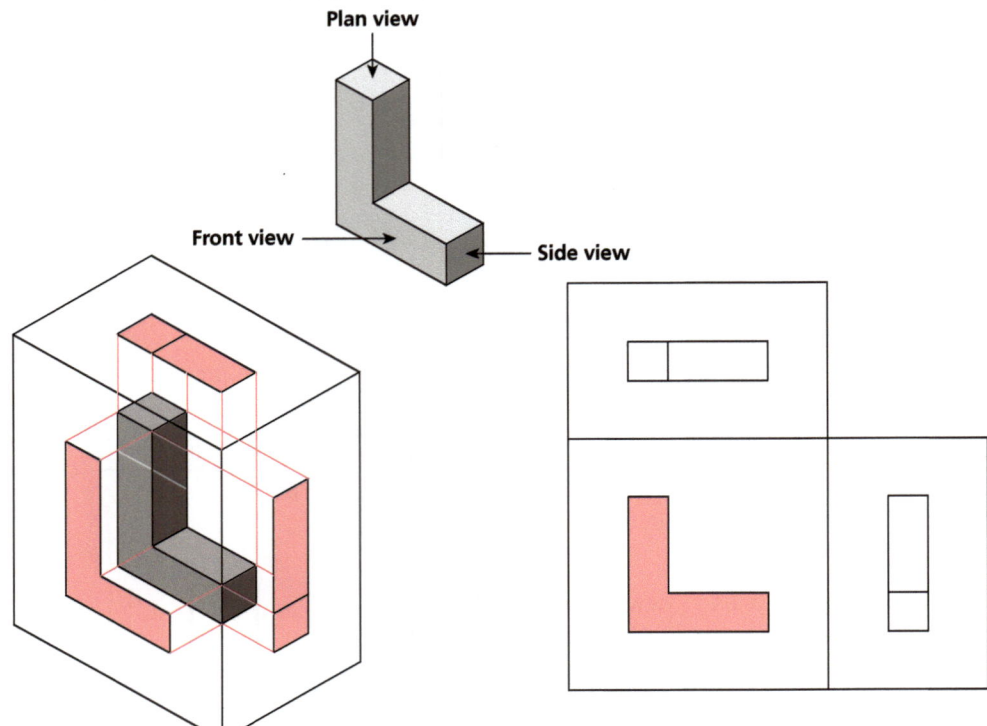

▲ *The 3D object showing all the views that will be displayed.*

Dimensions

Adding dimensions to your orthographic drawings is very important and must be completed accurately. The size and shape of your finished product are dependent on the dimensions you use on your drawing. To minimise any confusion when reading an orthographic drawing, you must use a standardised way of dimensioning (BSI 8888:2017). Here are a few simple rules:

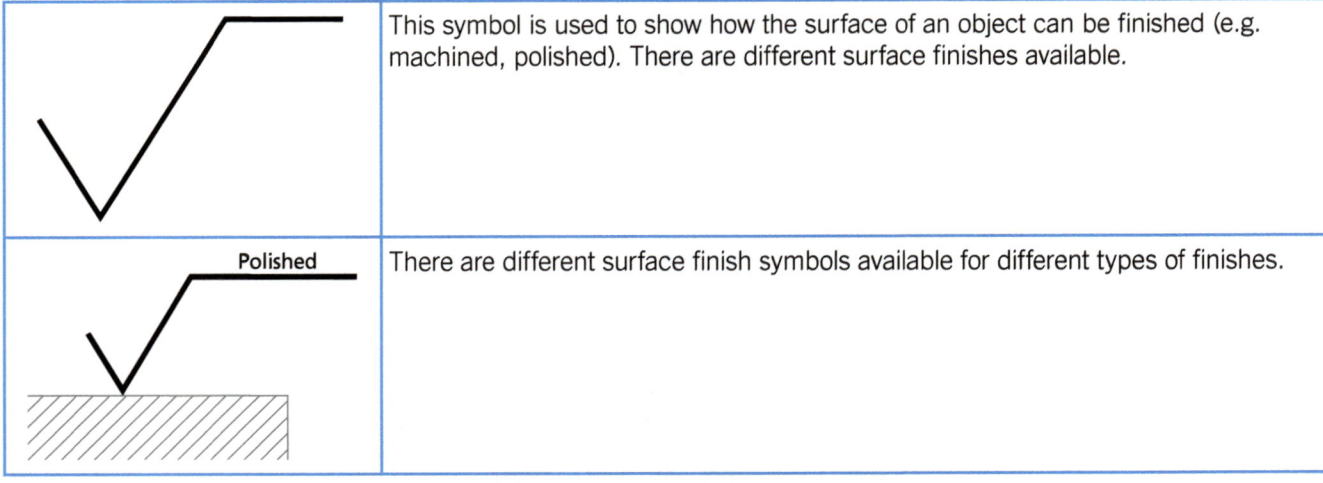

1 Understanding engineering drawings

	All dimensions must be kept to a minimum. You can dimension to the left or right; you can dimension above or below. Arrowheads must be a solid block. Numbers must be on the line (not in it or below it).
	If a measurement is 9 mm or less, the dimension number must be outside the extended lines.
	Radius measurements are shown with an R.
	Dimensioning angles should be drawn like this (use the end of the line as the centre point of a circle to draw the dimension line as an arc).
	Circles can be dimensioned with a radius (R) or a diameter symbol (Ø).
	This symbol is used for showing the depth of part of an object. For example, when looking at an object with holes that have been drilled, this symbol can show the depth of the drilled holes.
CBORE	CBORE stands for **counterbore**. This is a flat-bottomed hole created for a fastener (e.g. a bolt) to sit flush or below a surface. It is used in conjunction with a smaller hole. The CBORE symbol is used with the diameter symbol to show the different dimensions.
CSINK	CSINK stands for **countersink**. This is a conical hole created for a fastener (e.g. a screwhead) to sit flush or below a surface. It is used in conjunction with a smaller hole. The CSINK symbol is used with the diameter symbol to show the different dimensions. You can also show the combined angle of the countersink.
M8	A code is used to indicate thread size. In the example of M8: M = metric, 8 = 8 mm. Therefore, in this example, an 8 mm thread is needed.

Lines

Many different types of line are used when constructing an engineering drawing. Due to the sheer amount and variety of lines used, specific lines have been created to show specific things or that have a specific job. The figure below shows some common examples that conform to BSI 8888:2017 and details what they are used for.

▲ *Different types of line used in constructing engineering drawings*

Construction lines and weighted lines

Construction lines are very thin/faint lines used to construct the shapes you are drawing. They are used to tell you where the position of each object is.

Weighted lines are thicker/darker lines and are used to define or pick out the actual object.

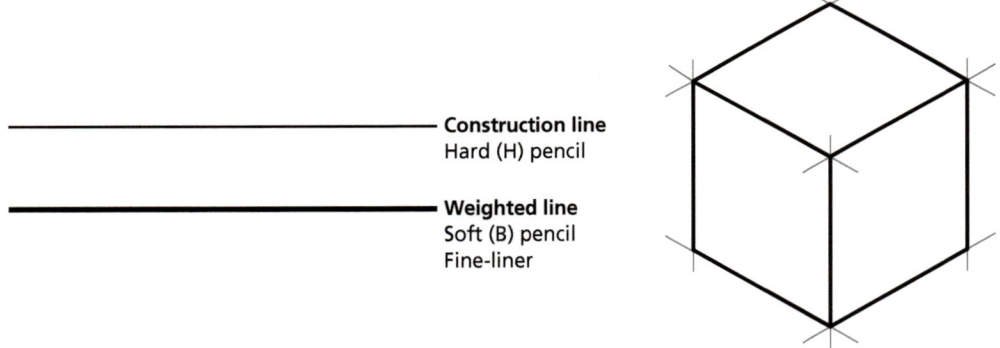

▲ *Different pencils are used for construction lines and weighted lines to give the different thicknesses required.*

Centre lines

Centre lines are used to show the centre point of a round object.

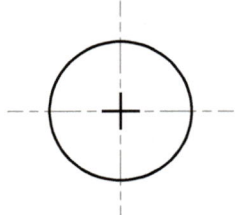 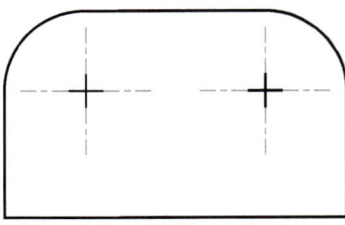

▲ *Centre lines show the centre point of circular objects.*

1 Understanding engineering drawings

Hidden detail lines
Quite often on an orthographic drawing, you will find objects with details that will be hidden when showing a certain view (e.g. plan view, front view, side view etc.). This hidden detail must be shown with dashed lines. In the drawing on the right, the hole in the object has been shown on the front and plan views with dashed lines.

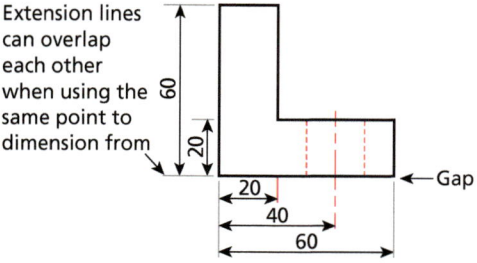

▲ Dashed lines are used to show hidden details.

Extension lines
The **extension lines** are those used in dimensioning and they define the area that is being dimensioned. There should be a small, consistent gap between the object you are dimensioning and the extension lines.

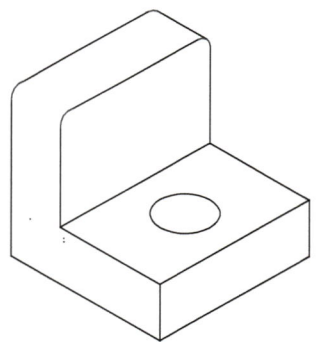

▲ Extension lines define the area being dimensioned.

> **Key Term**
>
> **Extension lines** These lines show the extent of the area that is being dimensioned. They are also known as lead lines.

Cut planes/section lines
On some views (plan, front, side), you may find a section line or cut plane line. This will coincide with a sectional drawing on the same page (see below).

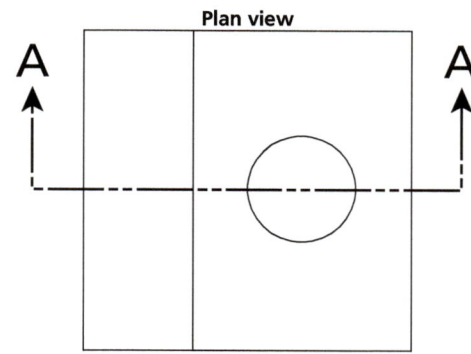

▲ Isometric view of an object

▲ Section line: the direction of the arrows shows what part of the object is being shown in the sectional drawing.

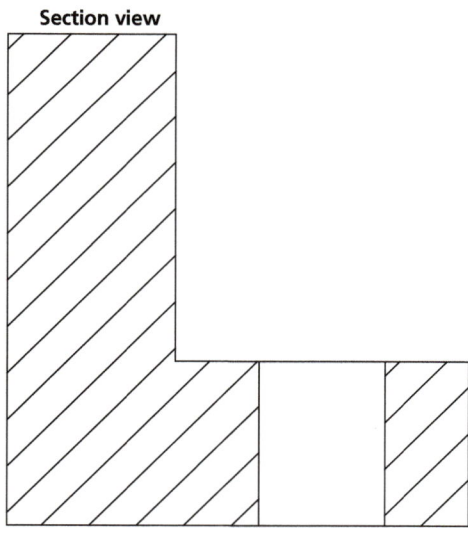

▲ Sectional drawing showing the part that is displayed. The hatching shows where the part has been cut.

Offset section lines
Offset section lines are similar to section lines but can deviate from a straight line to include detail that may not be captured with a standard section line. Offset section lines can turn at 90°.

▲ Offset section line: the direction of the arrows shows what part of the object is being shown in the sectional drawing.

19

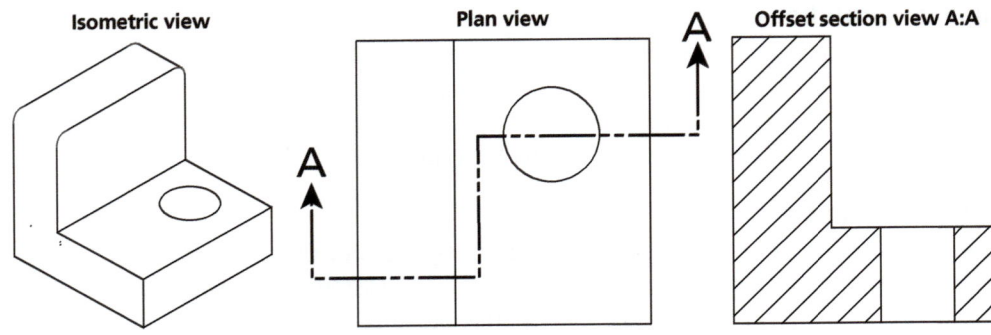

▲ Sectional drawing showing the part that is displayed. The hatching shows where the part has been cut.

> **Top tip**
>
> **Hatching rules:** when two different parts of the product meet in a section view, the hatching must (try) to go in opposite directions. All hatching must be evenly spaced (approximately 5 mm) and be at 45°.

The parts of a product that have been sectioned will show where they have been cut by the use of hatching.

> **Task 1.9**
>
> Draw a simple third-angle orthographic projection. Choose a 'view' and section it using section lines. Then draw the sectional view.

Tolerances

Tolerance or tolerance limits are measurement constraints that are commonly used to tell the manufacturer of a product how far off the stated dimensions the outcome can be. Tolerances usually have a plus or minus value and are shown like this:

+/– X mm (where X has a numerical value).

For example, if the tolerance for a long bridge is 100 metres plus or minus 5 metres, it would be written like this:

100 m (+/– 5 m)

The engineers that build the bridge know they can now build a bridge that is between 95 metres and 105 metres long.

> **Key Term**
>
> **Tolerance** An allowable amount of variation of a specified quantity, especially in the dimensions of a machine or part.

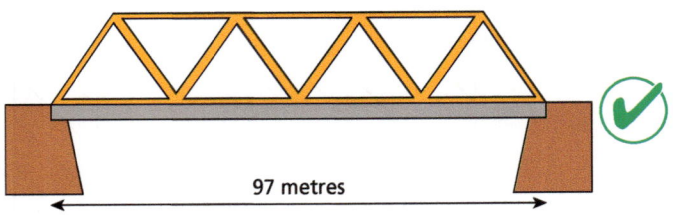

▲ The bridge in the figure has been built at 97 metres long. As the tolerance is 100 m (+/– 5 m), the bridge falls within the tolerance and is a good outcome.

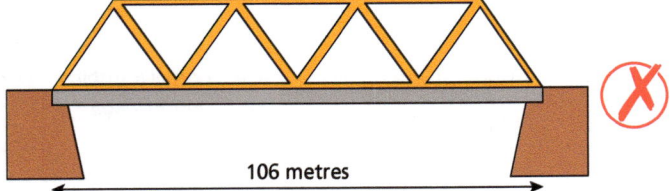

▲ The bridge in this figure has been built at 106 metres long. As the tolerance is 100 m (+/– 5 m), the bridge falls outside of the tolerance and is a poor outcome.

Border, parts list and title block

Before starting to construct an orthographic projection drawing, there are a number of tasks that need to be completed. These extra tasks allow for clarity of communication and also allow for further information to be added if needed.

Border

Borders are used to define the drawing and keep all the key information in one identifiable space. Nothing should be drawn or written outside of the border.

1 Understanding engineering drawings

Title block
A title block is a key part of any engineering drawing. It contains all different types of key information that may not be on the main drawing. The information in the title block can change from drawing to drawing depending on what is needed at the time. Typical information that can be found on drawings are:
- name or title of the part/project shown
- name of the author of the drawings
- scale of the drawing
- measurements of the dimensions
- date the drawing was created or updated
- number of drawing (e.g. drawing 1 of 3)
- materials used
- finishes used
- quantities
- any other information that may be relevant to the viewer.

Parts list
On an engineering drawing, a parts list is a table that can identify different parts within the drawing and offer further information on each part (similar to a title block). The parts list can be placed anywhere in the drawing but it must not interfere with any information that the drawing is communicating. Different types of information can be placed into a parts list depending on what is needed at the time. Some different types of information that can be included are:
- part number
- part quantities
- materials used
- finishes needed
- processes needed
- tools and/or machines needed for manufacturing
- any other information that may be relevant to the viewer.

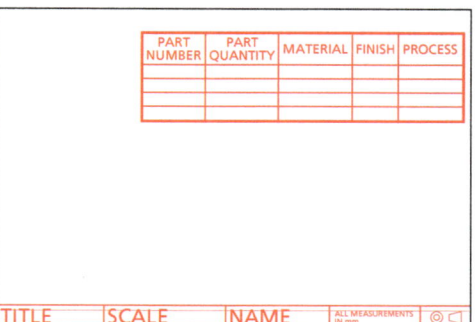

Projection lines and scale

Every orthographic drawing has projection lines. These projection lines are where you **project** the original view. They are very faint lines that help you to construct the drawing.

Each drawing also shows the **scale** to which it has been drawn. In general, it is impractical to draw something the size it actually is, so you often have to scale a drawing down or up. For instance, if your drawing is exactly the same size as the original item, this is written as 1:1; if your drawing is twice the size of the original, this is written as 2:1. The drawing below is half the actual size. The scale is written as 1:2.

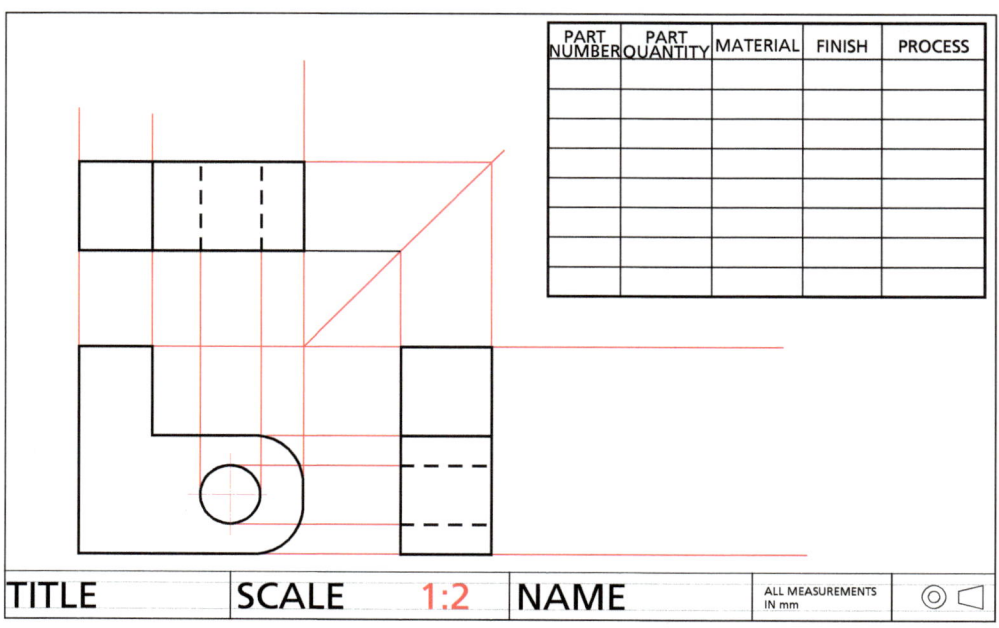

Task 1.10

Constructing a third-angle orthographic projection

To complete this task, you will need:
- 1 × drawing kit including a 45° set square, a 30° set square, a ruler, a compass, and H and B pencils
- 1 × A3 sheet of paper (landscape).

Follow the step-by-step guide on the next pages, using all the skills you have learnt so far. Try to be as accurate as possible.

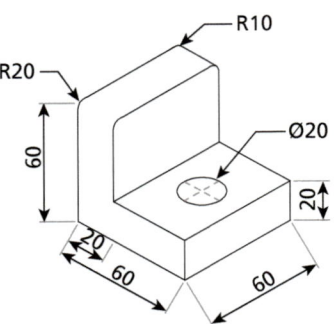

▲ This is the part/product you are going to be drawing. Try to keep the dimensions as accurate as possible when drawing your third-angle orthographic projection. All the units shown are in millimetres.

Step 1. Draw a neat border, a title block and an isometric view of the object you will be constructing. Don't forget the third angle symbol.

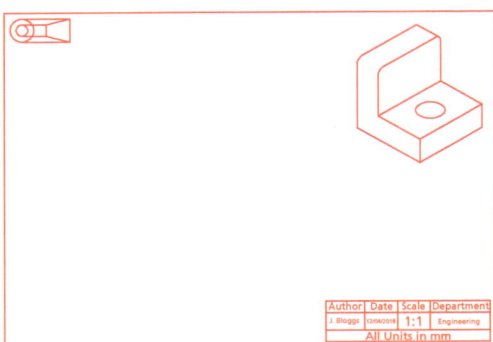

Step 2. Making sure you have the correct overall dimensions, draw a 2D crate and project the lines up (for the plan view) and across (for the side view) using construction lines.

(You do not have to draw in the dimensions yet.)

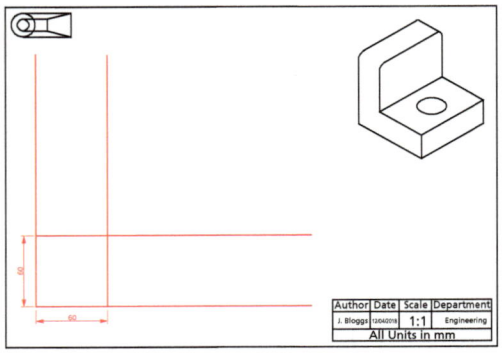

Step 3. Within your 2D crate, draw the front view of your isometric object, making sure the dimensions are accurate.

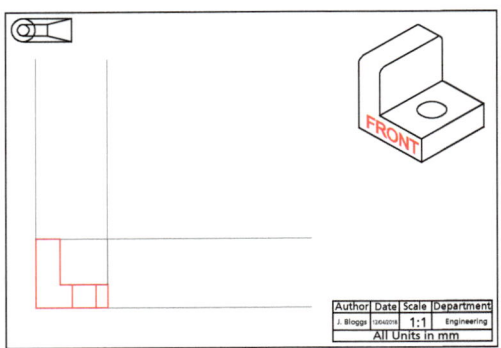

Step 4. Where there is detail on the front view (corners, holes, etc.) project those details as projection lines up (plan view) and across (side view). Using your 45° set square, draw a 45° line from the corner of your 2D crate.

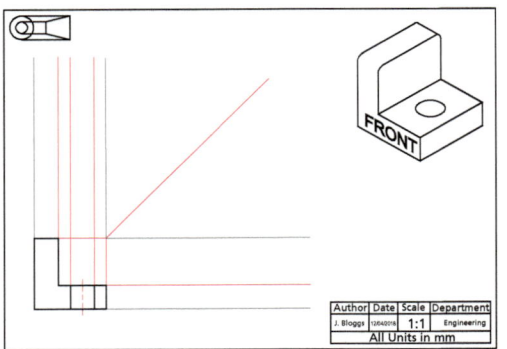

Step 5. Draw in your plan view, making sure the dimensions are correct.

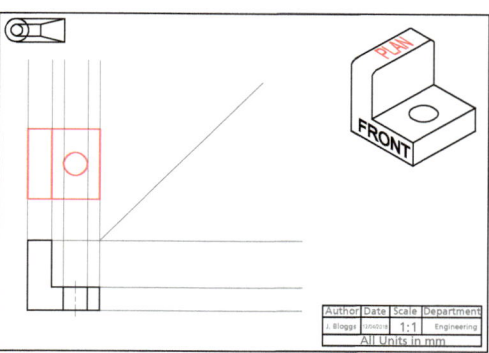

Step 6. Where there is detail on the plan view (corners, holes, etc.) project those details as projection lines across to the 45° line. Then, where your projection lines intersect the 45° line, project the lines down to create the side view automatically.

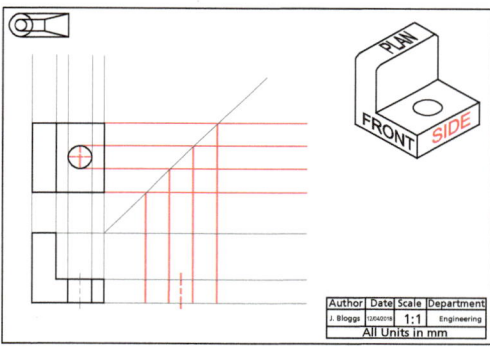

Step 7. Complete your side view adding any further details (e.g. radius, centre lines, etc.).

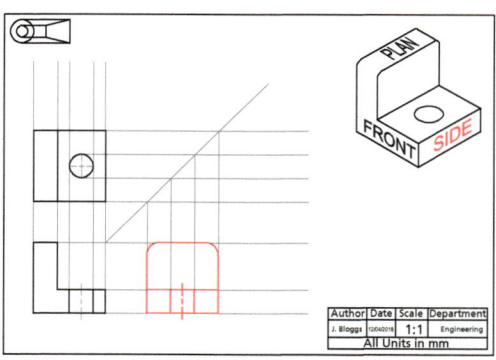

Step 8. Use weighted lines to pick out the three views and any other required detail (centre lines, hidden detail, etc.).

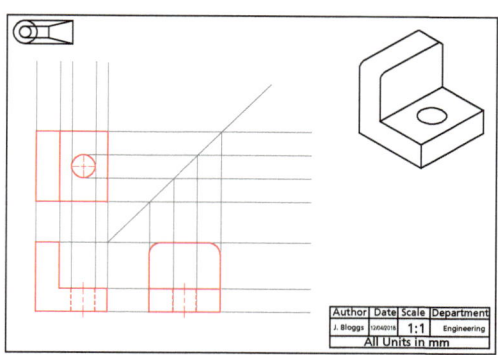

Step 9. Dimension your third-angle orthographic projection drawing using the guidelines on how to dimension accurately (see page 16). Add a parts list, including any details that you think a third person would need to be able to manufacture your part/product (quantity, materials, etc.).

Dimensions and the parts list can be seen in greater detail on the third-angle orthographic projection on page 24.

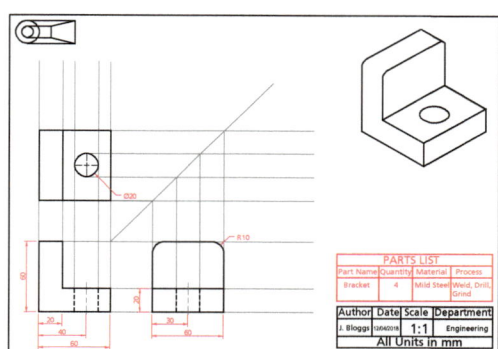

Step 10. Add in any extra details you think may be needed (e.g. sectional view; check page 19).

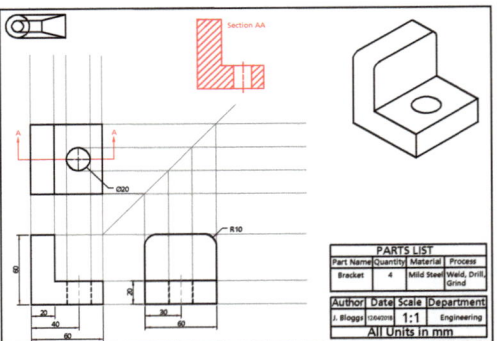

Step 11. Check your completed drawing for any mistakes. Finished!

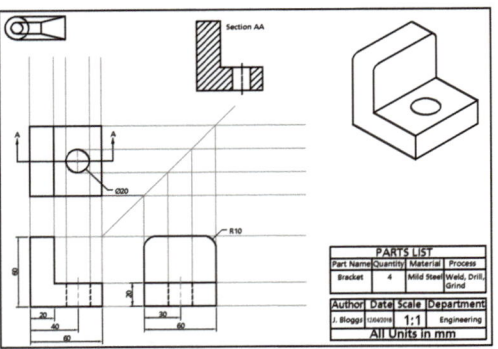

Task 1.11

In your notebook, list the items circled A–G shown on the third-angle orthographic drawing below. For example, A: The drawing is in _____ angle orthographic projection.

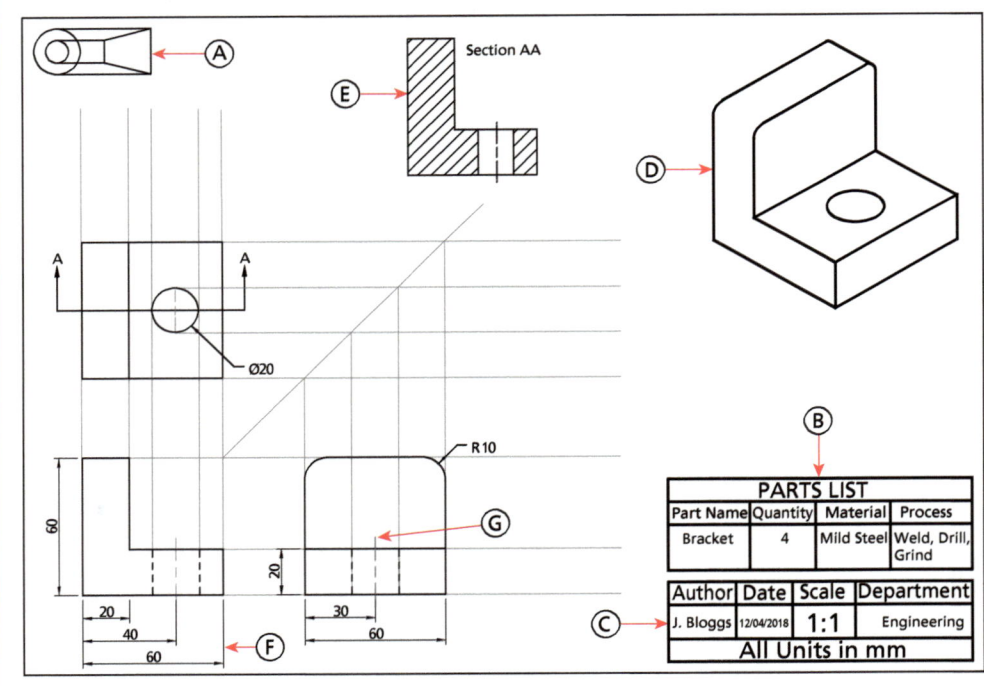

Task 1.12

To complete this task, you will need:
- 1 × drawing kit including a 45° set square, a 30° set square, a ruler, a compass, and H and B pencils
- 1 × A3 sheet of paper (landscape).

Using the knowledge you have gained in this chapter, construct an accurate third-angle orthographic projection of the object shown below.

Success criteria

Your drawing must:
- be dimensionally accurate
- use construction lines and weighted lines
- be dimensioned correctly
- include a parts list
- include a title block
- include a section view
- include any other standard conventions (e.g. third angle symbol, hidden details).

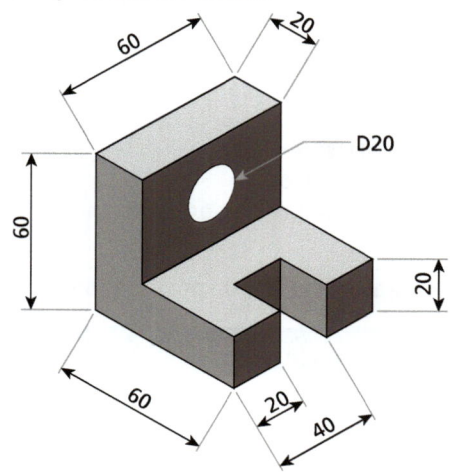

Top tip

When you are finished and the drawing is complete, ask yourself this question: Could a factory manufacture this product based purely on this drawing?

Top tip

Take your time. Focus on accuracy. If you are unsure, go back and check the lessons in this chapter.

2 Planning manufacturing

In this chapter you are going to:
- identify standardised material stock-forms
- create an accurate plan for engineering projects
- identify potential issues and barriers in engineering projects and create a contingency plan to accommodate these potential issues.

This chapter will cover the following areas of the WJEC specification:

Unit 1 Manufacturing engineering products: Unit 1.2 Planning operations		
• 1.2.1 Identifying materials	• 1.2.4 Planning and sequencing	• 1.2.5 Contingency planning

Stock-forms

Introduction

Before an engineer can start to produce an engineered product, they need to create a plan that will allow them to gain an effective outcome within the timeframe provided. To create an effective plan, they must not only consider timeframes, but what materials are required to create the outcome. To ensure a successful outcome, engineers must also identify any potential barriers and issues that could occur during the duration of the project.

You will be provided with a **brief** at the start of your project. A brief, also known as a **design brief**, is information or instructions provided to an engineer by a client or customer. The brief will ask the engineer to solve a problem or create a new design solution and will contain the information needed by an engineer to get started on the project. Different briefs can provide varying levels of information.

Identifying materials

Introduction

When an engineer has received the brief and created a solution to the context or problem, they need to start to identify the most suitable materials for their design from a range of available types.

Once materials have been extracted and refined to make them more usable, they are formed into various shapes and sizes. Rather than purchasing materials to exact measurements, engineers usually source and purchase materials in standardised shapes and sizes, called **stock-forms**. The benefit of standardised stock-forms is that engineers know exactly the shapes and sizes that are available for each material and they can therefore modify and improve their designs to suit the available stock-forms. Standardised stock-forms allow an engineer to place a price on a form of material, making it easier to cost projects.

Key Term

Stock-form The standardised size and shape in which a material is available from suppliers.

2 Planning manufacturing

Metal stock-forms

When engineers use metals, they need to know in what forms they can be supplied. A knowledge of the shapes available is important so engineers can ensure the correct shape is ordered for the project being undertaken and can use the correct terminology when ordering the materials.

> **Top tip**
> Metal stock-forms can come in a range of lengths.

Sections

A common metal stock-form is known as a **section**. You can work out the type of metal section by looking at it in cross-section and identifying what two-dimensional shape you can see. If you cut a square bar through the middle, for example, you will see a square cross-section. This 2D shape would show the engineer what stock-form is being used.

The following figures show examples of some metal sections and how they are supplied. The dimensions of the cross-section can vary. The metal sections can be extruded to the desired lengths.

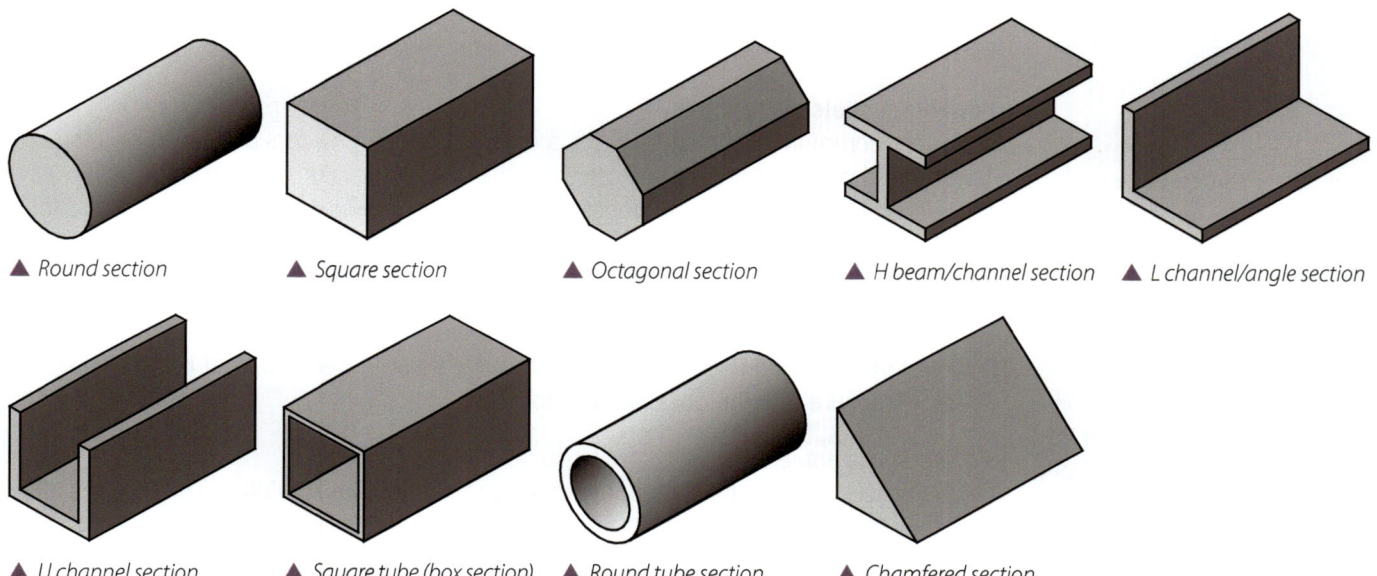

▲ Round section   ▲ Square section   ▲ Octagonal section   ▲ H beam/channel section   ▲ L channel/angle section

▲ U channel section   ▲ Square tube (box section)   ▲ Round tube section   ▲ Chamfered section

Extrusions

Metal stock-forms are also available in **extrusions**. An extrusion is effectively the same as a section – they are also cross-sections available in large lengths – however they come in far more complicated shapes. Due to the complex nature of extrusions, they are not very common and an engineer would likely request this stock-form for a particular project.

▲ Metal extrusions

Sheets

Metal sheets are a very common stock-form and are often known as **sheet material**. They are found in many useful household products such as washing machines, dishwashers, fire extinguishers and PC exteriors. Metal sheets are measured in **gauges**, which refers to the thickness of the material.

▲ Metal sheets are commonly used in such products as washing machines and fire extinguishers.

Plastic stock-forms

Plastics are a common material used in industry. Just like metals, they come in a range of shapes and sizes called stock-forms.

Sheets

Plastic sheets are flexible, however as they often come in thicknesses of over 1 mm, they are too thick to be rolled. Sheets are measured by thickness, width and length, so are available in a large range of sizes. Sheets are commonly used in industry and schools and can be used for processes such as vacuum forming (moulding), signs and PPE.

Film

Plastic film stock-form is thin enough to be available in very large rolls. It is widely used in industry for products such as packaging and coverings.

Sections/extrusions

As with metals, plastics can be ordered and purchased in sections and extrusions. Plastic sections and extrusions are often used on table edges and trunking.

▲ *Plastic sections/extrusions*

Powders, pellets and granules

The more common plastic stock-forms are granules, fine powders and small pellets. These plastics are measured and are therefore costed by volume and weight. They are used in different moulding processes to create the various shapes on many plastic products in common use.

Powders

Plastic powders are mainly used for **compression moulding**. This is where the powder is forced into a mould under high pressure to create a desired shape. It is then slowly heated to ensure that the powders fuse together.

▲ *Plastic pellets come in a multitude of colours.*

Pellets and granules

Plastic pellets and granules are commonly used in **injection moulding**. They are heated until they turn into a liquid and this is then formed into or around a mould to create the shape required.

Wooden stock-forms

Top tip

Wooden stock-forms can come in various lengths.

Timber can be referred to as 'rough cut' once it has been cut at a sawmill. **Rough-cut timber** can have many applications such as for building work and garden fencing. In general, timber sold from merchants and DIY stores has been smoothed and the edges **planed** or sanded. As a result, the price of that piece of timber will be more expensive than that of rough-cut timber due to the extra costs from the smoothing process. Such timber can either be **planed both sides** (PBS) or **planed all round** (PAR) (sometimes called planed square edge (PSE)). Planed timber is more desirable for interior projects, whereas rough-cut timber, due to cost, is more desirable for exterior products and projects. Hardwood and softwood can be bought in many different stock-forms, which are noted below.

Regular sections

Regular sections refer to the proportional dimensions of the timber. These are the most common sizes of timber that can be picked up in most DIY stores and timber merchants.

Mouldings

Moulding stock-forms are sections of timber with a decorative pattern that has been cut using a spindle moulder (a rotary cutting tool). They can be used for skirting boards and dado rails.

Dowel

Dowel is a cylindrical rod that is often used to align or fasten pieces of timber together. Dowel is measured in diameter and length and is available in a range of sizes, with common diameters ranging from 6 mm to 40 mm.

▲ Dowel comes in a range of diameters and lengths.

Sheets/boards: full board and half board

Boards or sheets are available to purchase in a range of materials, including plywood and MDF. Full boards usually have the dimensions of 1220 × 2440 mm. Half boards are available that are half the size of a full board. Both full and half boards are available in varying thicknesses.

▲ Full and half boards can come in a range of thicknesses.

Task 2.1

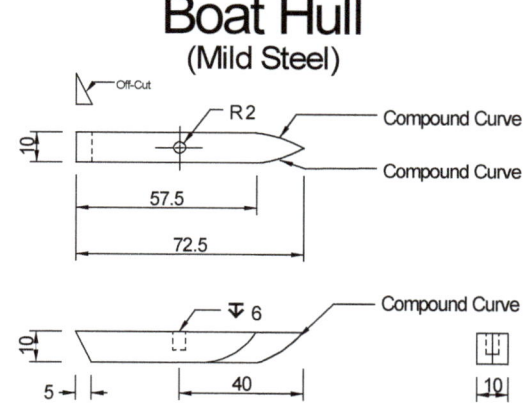

Boat Hull (Mild Steel)

Compound Curve, Compound Curve, Compound Curve
Off-Cut, R2, 10, 57.5, 72.5, ▽6, 5, 40, 10

Finishing:
After filing, the Boat Hull will be finished with a combination of Emery Cloth and buffed with the polishing machine.

A JIG must be created to hold the workpiece when using the Polishing Machine.

The steel 'off-cut' can be used as a 'fin' on the rear of the boat for aesthetic purposes if wanted/needed.

Tolerance = 1mm

Sail (Aluminium)

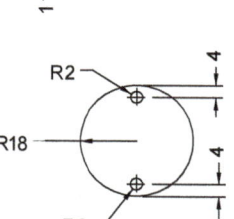

1, R2, 4, R18, 4, R2

Finishing:
After drilling:

The Aluminium discs will be 'Planished' using Planishing Hammers to create a dimpled effect on the sail.

The sail will then be curved to allow it to fit over the mast.

Tolerance = 1mm

Mast (Brass)

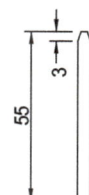

R2, 3, 55

Finishing:
The mast will be polished using a cloth and metal polish.

It will be assembled with the Boat Hull using adhesive.

The Sail will be attached using the pre-drilled holes.

Tolerance = 1mm

| TITLE: Metal Boat Project | SCALE: 1:1 | MATERIALS: Mild Steel, Brass, Aluminium | ALL MEASUREMENTS IN mm | 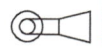 |

From the technical drawing provided in the figure, work out the suitable stock-forms for the following:
- the boat hull
- the boat sail
- the boat mast.

Planning and sequencing

Introduction

Before engineers can start a project, they need to create an effective plan. This is best practice to ensure that they are able to stay within the project budget and complete the project or product within the set timescale.

The following sequence shows the usual steps involved in planning a project:

Brief > Interpreting technical information (orthographic drawings) > Creating a cutting list > Creating job sheets and sequencing charts > Creating a Gantt chart (time management) > Creating a contingency plan

Planning projects

Brief

Before an engineer can proceed with creating a project or product, they require a **design brief**. The table below summarises what a design brief should cover. An engineer must be able to analyse a brief effectively, identifying the **key features**, and then propose a number of solutions that would satisfy or meet the brief.

Key Terms

Key features Relevant pieces of information.

Feasible Manageable or possible.

Target market The group of users you will be designing for.

Overview	Define your project and outline the scope of it.
Objectives	What are you attempting to achieve with this project? Is there a reason why it is important?
Target audience	Who are you hoping to target with this project/product? (The more specific you can be, the better.)
Budget	What is the overall budget for the project? How will this be broken down and spent?
Timeline and deliverables	For each deliverable, outline the date of delivery and a description.

The following is an example of a written design brief. This is typical of the series of statements that an engineer would receive from clients or colleagues.

> ### Example design brief for a new type of cycle
>
> A large manufacturer of bicycles and bicycle add-ons is looking to produce a new type of cycle to be used specifically for high-street shopping. The solution could be a bicycle, tricycle or any cycle combination that would be best for the shopping task. The cycle solution would need to be made from materials that the manufacturer currently uses to ensure production of a new product is **feasible** (manageable). There must be sufficient space to transport shopping items (clothing, groceries etc.) as well as an option for security. The new cycle solution would need to be used by a range of people with different heights. The **target market** would also need to be able to afford to purchase the new cycle, so cost of manufacture would need to be considered.

2 Planning manufacturing

In this brief, there are a number of relevant pieces of information, or key features, that an engineer would need to consider when producing successful solutions. However, there are other pieces of information that may not be that useful when developing solutions. It can be helpful at this stage to rewrite the brief in such a way that the key features are easier to pick out. Different engineers will have different ways of doing this, but some methods used to summarise or pick out the key features include:
- highlighting features of a brief
- writing a **condensed** brief
- writing a bullet-point brief
- writing a prioritised bullet-point brief.

(This will be looked at in further detail in Chapter 6 – Designing engineering products, on page 105.)

Key Term

Condensed A condensed brief has been reduced in such a way that anything not required is taken out.

Planning

By scrutinising the brief, using one of the above methods, and identifying all the key information, engineers are now in a position to create an effective plan that will produce a successful outcome.

A good detailed plan should consist of the following stages:
- Identifying resources required such as materials, tools and equipment.
- Identifying any skills or training required for tools and equipment.
- Planning a sequence of tasks in a suitable order.
- Planning for contingencies and any limitations or constraints that may hinder the project.

Interpreting key information

Now that the engineers have the key information from the brief, they start with technical information from a page or digital file. Engineers have to interpret this information accurately and organise data in a way that is easy to use, follow and apply. By spending time organising this information before starting the production processes, engineers are less likely to make mistakes and waste resources and are more likely to produce a working prototype to a high standard.

In this section you are going to look at how to interpret the information you are given as engineers and use it to plan the production of a product or solution. You will also look at how you could evaluate the outcome of the solutions.

Interpreting engineering drawings (orthographic projections)

When starting the process of producing a prototype (or any other production process), engineers will most likely have an orthographic projection (engineering drawing) with which to work. It might be that the engineer has produced the engineering drawing themselves or they might have been given it as part of a 'working group'.

The figure below shows an engineering drawing for several parts of a modern desk lamp. You can see from the drawing that the lamp is comprised of four different parts (circled in purple). However, if you look at the 'Parts List', you can see that the shade reflector is actually a **bought-in component** (circled in green) and would not need to be made by an engineer. So, already the relevant information is starting to be identified.

Key Term

Bought-in component A component purchased from another manufacturing plant that does not need to be produced/manufactured.

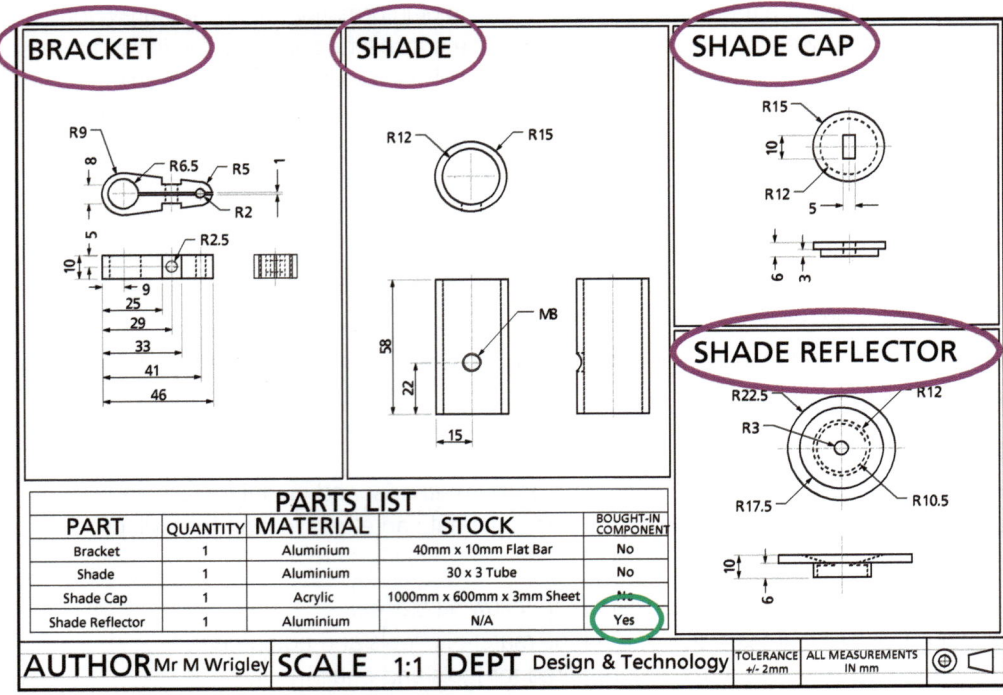

▲ This drawing shows the four listed parts (circled in purple) and how the shade reflector is a bought-in component (circled in green).

The next figure highlights the overall dimensions and sizes of the parts that need to be manufactured. By looking at the dimensions of the parts and the materials from which they are made, it is possible to identify how much material would need to be purchased and in what stock-form it can be purchased. The parts list also details how many of each part is needed (quantity), which would have an impact on the materials that need to be purchased.

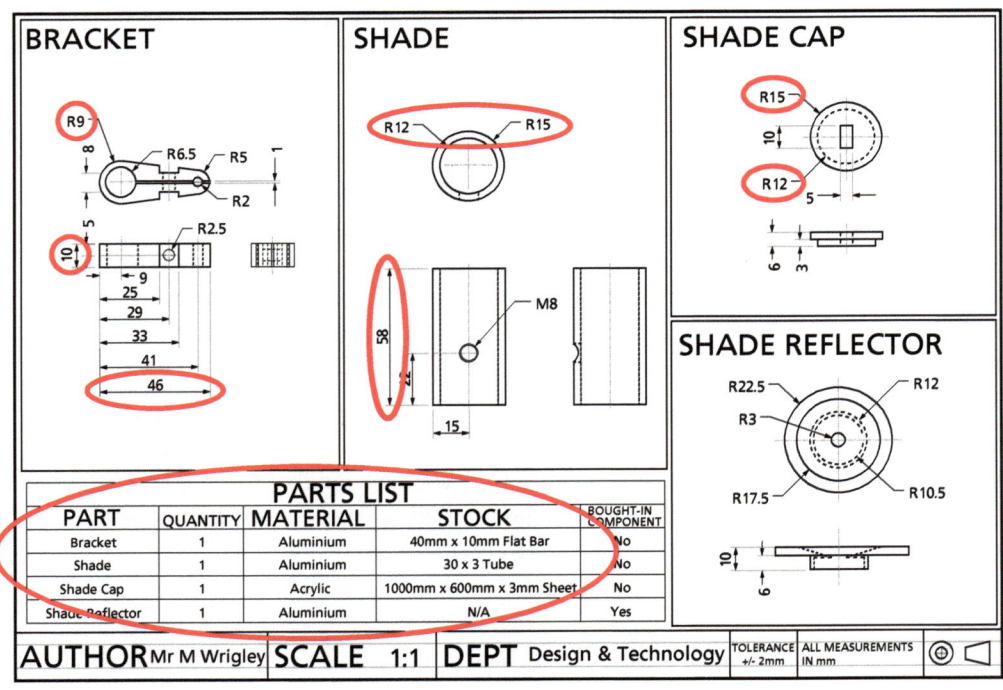

 **Top tip**

Looking at a parts list on any drawing gives you lots of information that benefits you when starting the organisation process of any new project.

▲ This drawing shows dimensions, materials and the amount of stock-form needed to complete the project.

Tolerance

The next figure highlights the amount of tolerance each part can have. This may have an impact on what workshop processes are chosen to complete the parts, as different processes have different levels of accuracy. For example, should a metal file or a milling machine be used to get a flat surface? What about a hacksaw or metal bandsaw to cut materials to size? Decisions made at this stage will have an impact on the quality of the outcome. Gathering as much information as possible and making 'informed' decisions will greatly benefit the project as a whole.

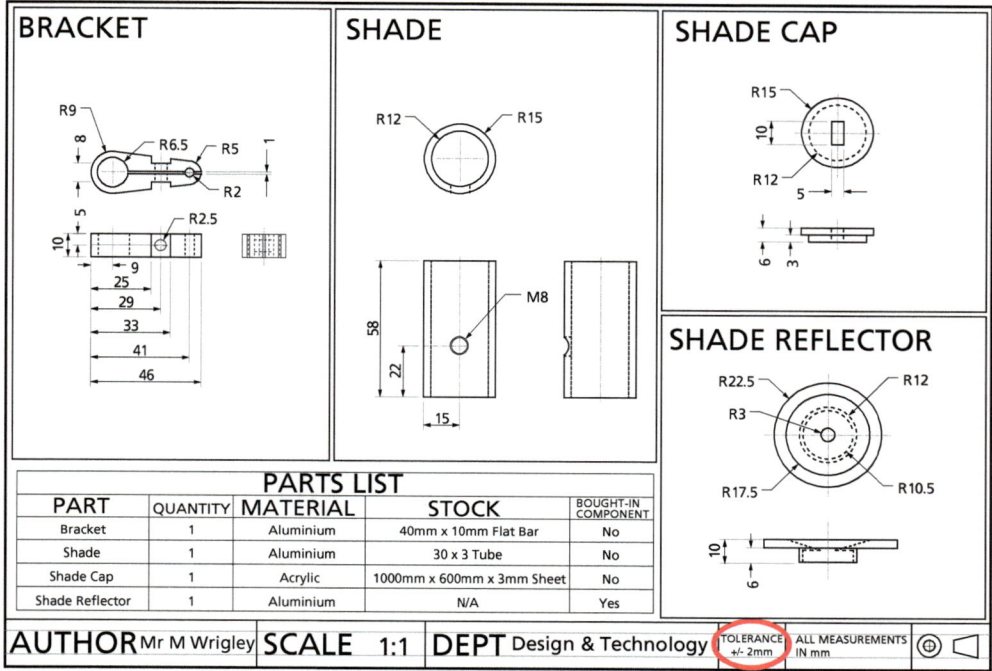

▲ The tolerance that each part can have is highlighted.

Creating cutting lists, job sheets and sequencing

Once you have identified the parts of an engineering drawing that will have the most impact on producing a prototype, you need to start looking at ways of organising this information so it can be used easily.

Cutting lists

A cutting list is a simple table showing how each part is to be cut and at what size (from the correct material/stock-form). The cutting list can be drawn up using the information gained from interpreting the engineering drawings. In the example shown below, notice how the information is organised to be read and used easily. The cutting list also includes details of the tools and process needed to cut the material to size.

Cutting list				
Part	Material	Stock-form	Cut to size	Tools/equipment needed
Bracket	Aluminium	50 mm × 10 mm flat bar	20 mm	Hacksaw/metal bandsaw
Shade	Aluminium	30 mm × 24 mm round tube	60 mm	Hacksaw/metal bandsaw
Shade cap (top)	Acrylic	3 mm sheet	R30 mm	Laser-cutter/bandsaw
Shade cap (bottom)	Acrylic	3 mm sheet	R30 mm	Laser-cutter/bandsaw

WJEC Level 1/2 Vocational Award Engineering (Technical Award)

Top tip

Another benefit of producing a cutting list is that it allows you to start thinking about what tools, equipment and processes you are going to need to produce your prototype.

The information in this table allows you to start sourcing the correct materials and begin the process of cutting them to size.

Job sheets and sequencing

A job sheet is similar to a cutting list as it can come in a 'chart' format and has information that can be used to help create prototypes efficiently and accurately. Job sheets, however, are designed to show what 'jobs' need to be completed for the project and also in what order they need to be completed. To create a part (from the engineering drawing) you would need to access the correct material, cut it to the correct size and then go through a series of processes that use tools and equipment to make sure the part is shaped and completed to the correct dimensions (that fall within tolerance). A job sheet can list all the tools and operations needed to complete the part successfully, as well as including other relevant areas such as risk, time and quality control.

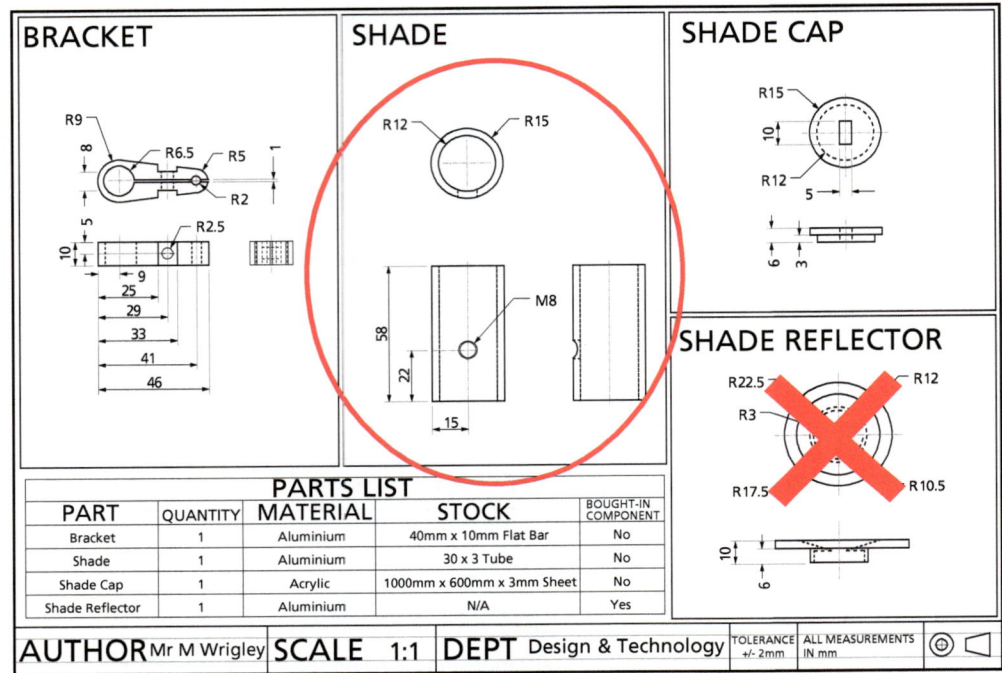

Top tip

To identify what processes are needed to complete a job sheet, you can go back to your engineering drawing and look at the orthographic projections of the part.

▲ *By focusing on one part (in this case, the shade) it is possible to identify the processes that need to be completed.*

By looking at the part 'SHADE' in the engineering drawing above, you can identify that to create the shade, the aluminium tube would need to:

1. be marked-out
2. be cut
3. be centre punched
4. be drilled
5. have a thread tapped.

To perform these tasks, you will need:

1. steel rule/Vernier calliper, scriber, engineer's blue, V-blocks
2. metal vice, hacksaw/metal bandsaw
3. hammer (ball-pein), centre punch, V-blocks
4. pillar drill, 7 mm HSS twist-drill bit
5. M8 tap, tap wrench, metal vice.

Once you have interpreted this information from the engineering drawing, you can put it into a format that will be useful when creating the prototype.

It is **very important** to note that the jobs and processes you list should be written in a workable **sequence.** This means that all processes should be completed in order. For example, when you look at the processes that have been identified for the shade, you can see that it needs to be drilled. Before the drilling takes place, however, you would need to make sure you are drilling in the right place. Therefore, marking-out, cutting and centre punching are three processes that would need to be completed *before* drilling takes place. Getting these tasks in the correct order is known as sequencing. Sequencing would also need to be applied correctly to the job sheet.

How you lay out your job sheet is up to you. The company you work for may have a standardised format that you work to. The main function of any good job sheet is to make sure you have quick and efficient access to the relevant information needed to complete the task successfully.

The following table is a job sheet that has been created for the shade part in the engineering drawing example. Look at how the chart has been created and what titles/columns have been added to give more information. This job sheet includes sequencing as well as risk assessing, quality control and time. The layout is easy to understand and use in a workshop environment.

Key Term

Sequencing Putting tasks into the correct order.

Top tip

Try to use a format that allows you to achieve the right priority.

Job sheet
Part: SHADE (aluminium round tube 30 mm × 24 mm)

Part	Material and stock-form	Process and sequence	Tools/equipment needed	Risk level	Time	Health and safety considerations	Quality control
SHADE	Aluminium round tube 30 mm × 24 mm	Step 1: Mark-out	Steel rule/Vernier calliper, scriber, engineer's blue, V-blocks	Low	10 mins	n/a	Check accuracy of marks
		Step 2: Cut	Metal vice, hacksaw/metal bandsaw	Medium	5 mins	Sharp hacksaw blade	
		Step 3: Centre punch	Hammer (ball-pein), centre punch, V-blocks	Low	5 mins	Strike centre punch squarely. Secure round tube in V-block	
		Step 4: Drill	Pillar drill, 7 mm HSS twist-drill bit	Medium-high	5 mins	Correct set up of pillar drill. Wear PPE	Check set up and drill **RPM**
		Step 5: Tap a thread	M8 tap, tap wrench, metal vice	Low	10 mins	Sharp teeth on tap	Keep tap vertical to ensure straight thread

Using data sheets

When creating a prototype in a workshop environment, you will more than likely be expected to use machinery to successfully and accurately create a prototype. You should also be tutored on the correct procedure for using the various machines that would include the correct set up of the machine, the correct way of using it, as well as what health and safety precautions you would need to incorporate.

Key Term

RPM Revolutions per minute. This is a measure of how fast a machine spins.

Top tip

When creating your job sheets, it is a good idea to have another column next to the process and sequence and tools/equipment columns with the correct speed setting or set-up settings for each process.

However, many machines come with interchangeable cutting tools (e.g. drill bits), with different tools needed for different materials. This is where you need to start using data sheets/charts that are industry standard guidelines or specifications set out by the machine manufacturer.

Using the correct settings and speeds for the machines will ensure:
- a good-quality finish
- a longer working life of the part/machine (no breaking)
- safer operation (user not being hurt).

By not using the correct data you could:
- destroy your work (have to start again)
- break parts or all of the machine
- injure yourself or others.

The following table is an example of a data chart for a centre lathe. It contains useful information such as what RPM is needed when using different materials at different sizes. Other useful information on RPM when performing specialist operations such as parting and knurling is shown in the Top tip.

Top tip

Knurling and parting operations should only be run at 100 RPM maximum.

Guideline RPM for a centre lathe					
Material diameter		Material			
Inches	Millimetres	Aluminium	Brass	Mild steel	Stainless steel
½	12.7	1400	1200	1000	600
1	25.4	700	600	500	300
1½	38	500	400	300	200
2	50.8	350	300	250	150
2½	63.5	280	250	220	120
3	76	225	190	160	100

Time management

Gantt charts

Another big resource for many engineers to consider, whether working for a company or even trying to complete a prototype in school or college, is time.

Time (along with cost) is one of the main **constraints** engineers have when starting projects. Being able to organise time effectively is of great benefit when trying to manage projects. Having the ability to organise time effectively for projects means time is not wasted, money is saved and deadlines will be met.

Key Term

Constraint A limitation in a certain situation.

Gantt charts (invented by Henry Gantt, 1861–1919) are widely used by engineers and in industry to organise time effectively. The time allocations for the various steps in the production of the shade were included in the job sheet. However, having a chart that organises time, which can be scanned and understood quickly, is a very effective way of managing time.

The following table is an example of a Gantt chart. This chart has been drawn up to break down the time needed for the manufacture of the shade part of our desk lamp.

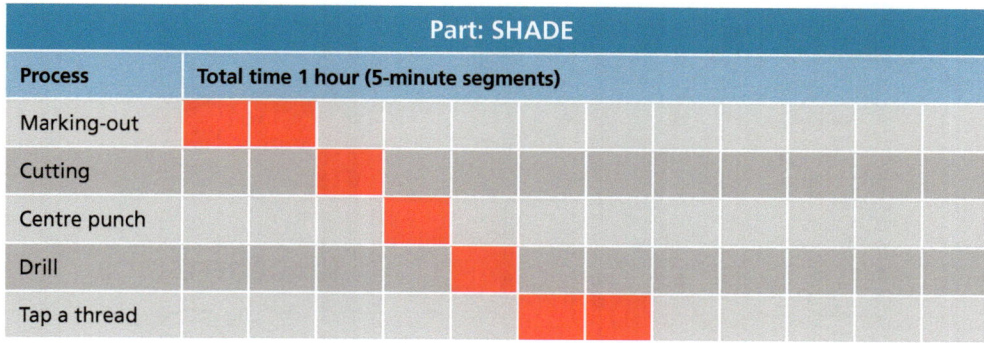

▲ An example of a simple Gantt chart

A quick look at this Gantt chart shows that the processes for the shade part of the lamp will take a total of 35 minutes, with marking-out and tapping a thread taking the longest time to perform. With this information in mind, you can start identifying areas in the workshop or manufacturing facility that could have bottlenecks and slow production. This information will help you organise your time more effectively.

On larger projects, when manufacturing in industry, a more complicated Gantt chart may be required to manage the timings of multiple tasks being performed by larger teams of people that could be timed over weeks and months. The following is an example of a Gantt chart that could be used for more complicated projects.

▲ An example of a Gantt chart for a project involving many stages.

Task 2.2

Now you understand the importance of organising relevant information, and are able to use it quickly and efficiently, you can try producing your own job sheet.

Look at the following engineering drawing. Create your own job sheet for the part 'BRACKET'.

You must include the following:
- titles
- materials or stock-form
- sequence of tasks (jobs)
- equipment needed
- time.

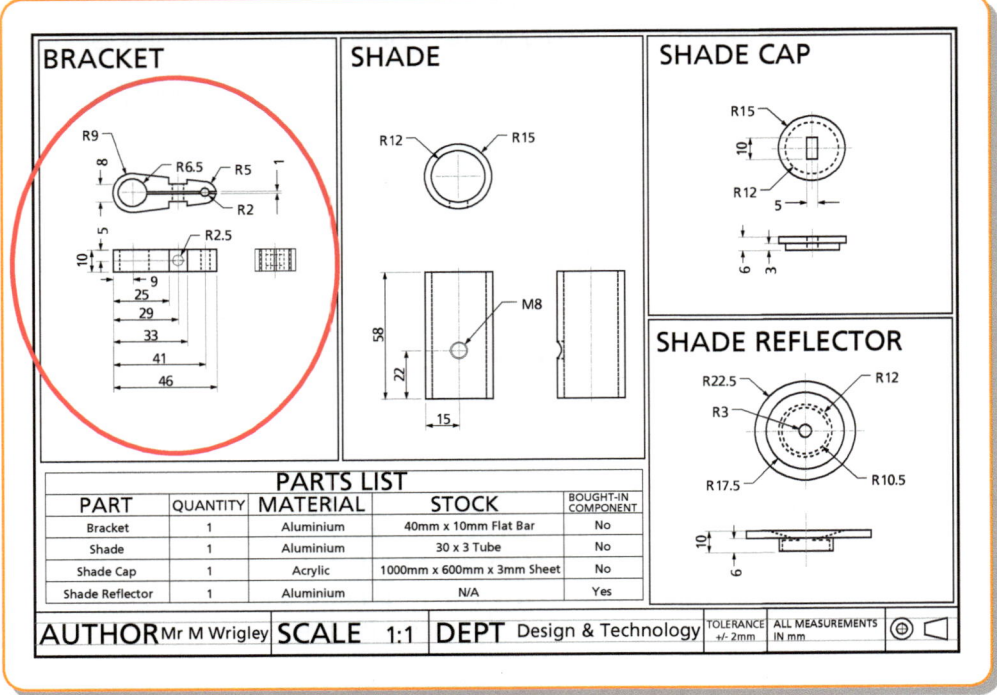

Contingency planning

When planning a project, engineers must also consider that even the best plans do not always go the way you expect. It is important, therefore, to draw up a **contingency plan**. In a contingency plan, engineers attempt to foresee any potential problems or issues that could occur when creating their product or project. Some of the potential issues could be out of the engineer's hands; however solutions still need to be provided. Engineers look to provide solutions to these problems before they occur to ensure the project can progress and deadlines are met.

Some possible issues that may need to be considered include:
- illness
- school or college closures
- unavailability of/limited tools and equipment/incorrect or faulty tools
- equipment and machinery failure
- shortages of materials or materials being unavailable.

Top tip
Depending on your facilities and experience, you may encounter further problems which are not on the list.

Illness
Unfortunately, through no fault of your own, you could become ill and, therefore, unable to attend school or college. If this happens, you will need to look after yourself and concentrate on recovering; your project work will have to stop temporarily. To ensure any bouts of illness do not stall the progress of your project, an additional few hours could be included in your initial plan or Gantt chart to allow for such circumstances. This will accommodate any 'mop-up sessions' and should enable the project to be completed within the timeframe set.

School or college closures
It is possible that your school or college may be closed for inset days or for various reasons out of everyone's control, in which event you may be unable to use the workshop facilities to continue with the physical making of your project. To ensure timescales are still met, you should ensure that a plan is in place to retain access to your written work, plans and working drawings via the internet or various drives.

For example, gaining a further understanding of the working drawings, continuing to create and develop parts and cutting lists or completing an evaluation on parts of your project that you have already completed, will allow you to improve aspects of your work when you return to school or college.

In a modern society, laptops and computers are readily accessible, either at home or at local libraries, and a lot of CAD software is cloud-based, meaning that it does not need to be installed on your computer. If your school or college is closed, you should still be able to use CAD software to create models, CNC/CAM manufacturing plans and parts lists, so there is potential to continue with your project via the CAD/CAM route.

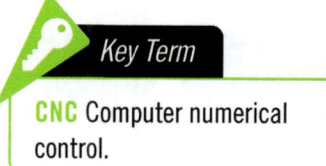

CNC Computer numerical control.

Taking these steps will allow you to make progress and prevent the project stalling.

Unavailability of/limited tools and equipment/incorrect or faulty tools

The key pieces of information that an engineer should refer to if tools become limited or unavailable are their cutting lists and Gantt charts. By reviewing these documents, engineers can identify whether tasks can be performed in an alternative order, rather than the sequence in which they are presented on the technical drawings. So, if certain tools or equipment are unavailable for the task that the engineer would like to complete, they are in a position to identify which tools *are* available and potentially link them to a suitable task that can be completed as an alternative option.

For example, if the pillar drill is unavailable due to maintenance or because it is being used by another engineer, how could the project continue? The engineer could use a hand-held drill to drill the material and an engineer's vice to keep the material securely and safely in place. It may be a more difficult process, and it may take longer, but by adopting an alternative method, the project can continue, the engineer is not wasting any time and they can keep to the timescale.

Equipment and machinery failure

There are times where machinery and key equipment are unavailable through servicing or failure. Again, contingencies need to be in place to ensure the project can continue. Similar to plans that you would put in place if there were limited tools available, you should consider a different approach to the task. For example, if your project requires the use of a milling machine but it is out of order, what tasks could you complete to gain the same outcome and prevent the project from stalling? A potential solution could be to use a hacksaw to cut your desired piece and a file to further enhance the accuracy. A milling machine will be more efficient and provide a higher-quality finish, but having potential solutions in place for possible issues with key machinery will allow your project to continue and deadlines to be met.

Shortages of materials or materials being unavailable

A technical drawing will specify the materials that would be most suitable for a particular project. Unfortunately, there can be occasions where that material is simply not available. It could be for any of the following reasons:
- **local availability:** demand in your local area for materials is high or there is a lack of available materials that can be sourced locally
- **material shortages:** there may be a shortage of the material you require for your project, therefore lead times could be high
- **high cost:** the cost of the material is too expensive and therefore not financially viable for the school or college.

If accessing materials becomes an issue, you should look at alternative materials that have the properties you need for your project, but that are available and affordable. The technical drawings tend to specify the most desirable materials to work with; however in many instances, there are alternatives that will work just as well.

3 Using engineering tools and equipment

In this chapter you are going to:
- accurately identify tools and equipment found in a workshop environment
- accurately state the function of tools and equipment used by engineers
- select the correct tools and equipment to perform specific tasks
- identify CAD and CAM
- discover why health and safety is needed in the workshop environment
- learn why personal protective equipment is important in health and safety
- learn about risk assessments
- understand health and safety signs (shapes and colours)
- learn about wearable protection
- learn how to use data sheets.

To achieve the objectives successfully, you should have access to a workshop environment with enough tools and equipment to demonstrate your engineering skills. (Your place of learning should provide these facilities.) This chapter does not look at all the equipment in great depth but it will give you an understanding and awareness of the tools available so you can identify them, and even specify equipment when a task needs to be completed. To gain further knowledge on how to use the equipment successfully, your tutor will be able to demonstrate far greater technical information in the correct environment.

This chapter will cover the following areas of the WJEC specification:

Unit 1 Manufacturing engineering products: Unit 1.2 Planning operations		
● 1.2.2 Equipment selection	● 1.2.3 Tool selection	
Unit 1 Manufacturing engineering products: Unit 1.3 Using engineering tools and equipment		
● 1.3.1 Using engineering tools	● 1.3.2 Using engineering equipment	● 1.3.3 Health and safety

Introduction

The technical knowledge you gain from this section is knowledge that you will be expected to demonstrate in a workshop environment for Unit 1. You will be given a series of orthographic drawings (working drawings) and you will be expected to produce the product shown in the drawings using the knowledge you gain from this and the following chapter. You will also be required to demonstrate knowledge of the equipment and processes in the workshop, along with knowledge and application of health and safety in a working environment. Your tutor will assess and grade you as part of the ongoing Unit 1 process.

3 Using engineering tools and equipment

Engineering equipment selection

Measuring

Steel rule
A steel rule is similar to a generic plastic ruler, but is more durable. Steel rules are also more accurate as they start directly from 0, whereas a standard ruler has a plastic edging, which makes it less accurate and less reliable. Steel rules have mm, cm and inches as the measurement options.

▲ Using a steel rule is more accurate than using a standard plastic ruler.

Measuring tape
A measuring tape is a flexible tool used for measuring length. It can be made from materials such as fibreglass, cloth, plastic, metal ribbon or metal strip. It is marked in both centimetres and inches.

▲ Measuring tapes can be made from a number of different materials.

Micrometers
The micrometer is another useful measuring device which allows for very accurate measurements. It is commonly used to measure external diameters of round section stock-forms and ball bearings.

▲ A micrometer allows for accurate measurements.

Callipers
Callipers are used for measuring various dimensions of a workpiece. Callipers can be 'set' to a specific measurement and then used to check the measurements while work on the workpiece is in progress. This saves time using steel rules and having to read the measurements numerous times to check accuracy, as well as offering consistency.

Vernier callipers
Vernier callipers are useful measuring tools that allow the user to measure external diameters, internal diameters and depths. They are very accurate and can commonly measure to 0.02mm accuracy (one fiftieth of a millimetre). Digital Vernier callipers are also available.

> **Top tip**
>
> Measuring tools can use metric measurements (mm, cm etc.) or imperial measurements (inch).

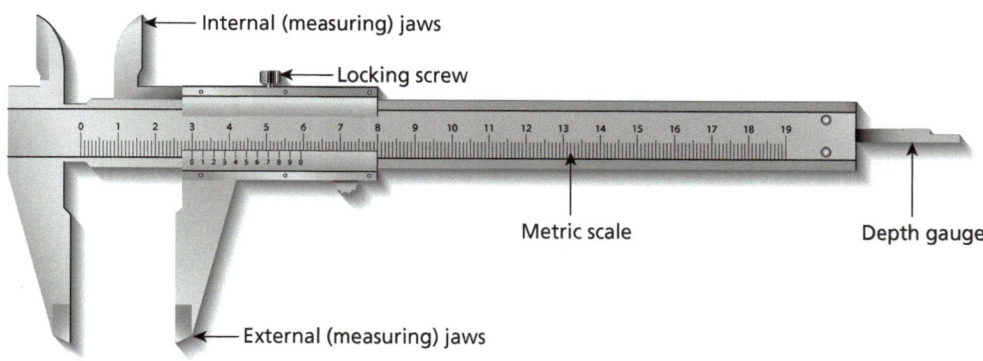

▲ A Vernier calliper

▲ Internal callipers can be used to check the internal diameter of a hole.

Internal callipers
Internal callipers are used to check the dimensions of an internal area while it is being worked on (mainly a hole). A good example would be checking the internal diameter of a hole while you are boring-out a pre-drilled hole on a centre lathe.

External callipers
External callipers are used to check the dimensions of an external area while it is being worked on. A good example would be checking the external diameter of a workpiece that you are 'turning down' on a centre lathe.

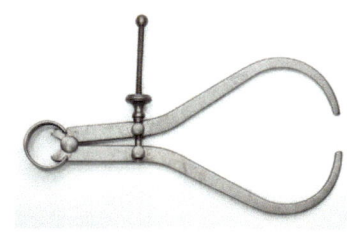

▲ External callipers check the dimensions of an external area.

41

Key Term

Scribe To mark out.

Odd-leg/Jenny callipers
Odd-leg callipers, also known as Jenny callipers, are used to **scribe** straight/parallel lines to the edge of a workpiece or section. They can also be used on circular sections of materials.

Combination square
A combination square is a measuring tool that can be used in metalworking and woodworking. It consists of a rule and an interchangeable head that can be attached to the rule. It is primarily used for ensuring the integrity of a 90° angle, measuring a 45° angle or measuring the centre of a circular object.

▲ A combination square is used in both metalworking and woodworking.

Centre finder
A centre finder is designed to quickly and accurately find the centre of round or square stock; this could be metal or wood. An engineer will lock the centre finder against the material stock and draw a line, then move around the circumference and draw another line. The centre is where the two lines intersect.

Angle finder
As the name states, an angle finder is a measuring device that measures and determines the angle of an object. This could be used to measure any form of material provided. The digital angle finder uses battery power to display results in a digital reading, while many other angle finders feature a dial and a needle to decipher an angle.

▲ A digital angle finder

Protractor
A protractor is a common tool that is used to measure angles of different materials. Most protractors measure angles in degrees (°). Radian-scale protractors measure angles in radians.

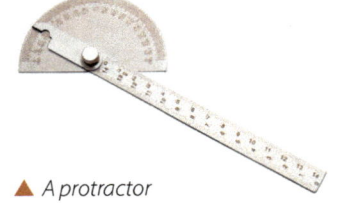

▲ A protractor

▲ A sliding T-bevel, used to construct angles

Sliding T-bevel
A sliding T-bevel is used for constructing and transferring angles. This tool is predominantly used for woodworking projects, but it can also be used on other materials, such as plastic and metals. This tool is ideal for when an angle of 90° is not possible and more complex angles are required.

Multimeters
A multimeter is an electronic device that is used to measure the current (A, amps), voltage (V, volts) and resistance (Ω, ohms) of a system or circuit. Some have a digital readout, others an analogue (dial) readout. They are used as a troubleshooting device to find faults within circuits and systems, as well as for testing circuits or systems when carrying out maintenance. You can even use them to test the power levels left in batteries.

Marking or scribing

Engineer's try square
The engineer's try square is a tool for scribing perpendicular lines (90°) on a section of material. The stock is placed alongside the workpiece section and the blade rests on the material at 90°. The engineer's try square slides easily up and down the section, making accurate and straight lines. It can also be used to measure whether your workpiece is even.

▲ A digital multimeter is used to measure current, voltage or resistance in a circuit.

3 Using engineering tools and equipment

Try square
A try square, or carpenter's try square, is used in the same way as an engineer's try square; however it is used for marking and checking 90° angles on pieces of wood. The square in the name refers to the 90° angle.

Mitre square
A mitre square is used in woodworking and metalworking for marking and checking angles other than 90°. Most mitre squares are for marking and checking a 45° angle and its supplementary angle, 135°.

▲ A mitre square can be used in both woodworking and metalworking.

▲ An engineer's try square (metalwork)

▲ A try square (woodwork)

> **Top tip**
> Any of the above 'square' tools should be pressed up against the material when being used to ensure accuracy.

Scriber
A scriber is a hand tool used for marking-out areas ready for machining/cutting/drilling, etc., on workpieces made from metal. The scriber is made from high-carbon steel and is hardened to make sure it can score the surface of the metal.

Dividers
Dividers work very similarly to compasses. Instead of having a pencil at one end though, the dividers have scribers at both ends. This enables you to scribe circles onto metallic surfaces.

▲ Dividers can scribe circles onto metallic surfaces.

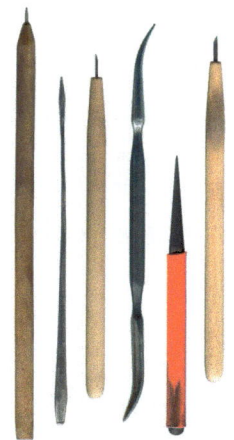
▲ Various scribers

Engineer's blue
Marking blue/engineer's blue is a liquid that can be painted onto a surface and then scribed through to create a thin line. The blue allows the engineer to see their marked line more clearly than if they scribed directly into the metal. The use of engineer's blue allows for a greater level of accuracy on metal-based projects.

Surface gauge
A surface gauge is a scriber attached to an adjustable stand that can also be magnetised. The surface gauge can move around a flat surface to scribe horizontal lines very accurately. It can also be used to check the accuracy of flat surfaces from processes such as milling and planing.

Surface plate
A surface plate, also known as a true plane surface, is a standardised piece of equipment that enables precise engineering measurements to be taken. It is often used in conjunction with a surface gauge to ensure exact measurements on the material.

▲ A centre punch is used prior to drilling.

Centre punch

A centre punch is used to mark-out the centre of a hole in readiness for drilling. The centre punch also creates a small crater that allows the drill bit to sit-in and bite as opposed to skating around the surface and possibly drilling in the wrong place. Centre punches are generally made from steel with a hardened tip. The centre punch is hit into the material with a hammer one or two times.

Marking gauge

A marking gauge is more commonly known as a scratch gauge. It is generally used in woodworking projects but it can also be used on metal projects. Its main purpose is to scratch and mark out accurate lines in readiness for cutting the material.

Marking knife

A marking knife is a tool used for marking workpieces accurately. It cuts a visible line, which can then be used to guide a hand saw or chisel. Marking knives are generally used when marking across the grain of the wood.

Bradawl

A bradawl is used to make indentations in wood in order to ease the insertion of a nail or screw. The blade is placed across the fibres of the wood, cutting them when pressure is applied. The bradawl is then twisted through 90°, which displaces the fibres creating a hole.

▲ A marking gauge or scratch gauge

▲ A marking knife

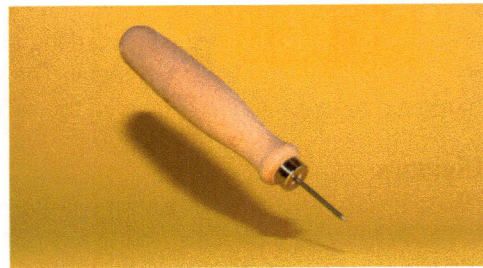

▲ A bradawl makes indentations in wood.

> **Key Term**
>
> **Jig** A workpiece-holding device that enables an operation such as drilling to be carried out repeatedly and safely.

Holding material

V-block

A V-block is essentially a **jig** used to hold round or cylindrical sections of metal or plastic while they are being marked-out (scribed), drilled or any other relevant operation. A V-block should have clamps attached that can screw down to hold the workpiece in place while it is being worked on. The V forms a 90° angle. V-blocks are generally sold in pairs so that they can be used for longer sections of metal.

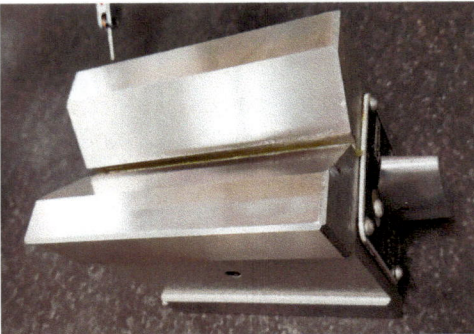

▲ A V-block holds cylindrical sections of material in place.

▲ A woodworking vice holds your wooden workpiece in place while you work on it.

Woodworking vice
A woodworking vice is an instrument that is designed to hold your piece of timber in place, so that you are able to mark, cut, sand or use a rasp safely and accurately. It is attached to a workbench. You should use a bench hook with this type of vice.

Hobby bench vice
The hobby bench vice is ideal for use with portable workbenches, the kitchen table or wherever a strong temporary fixing is required. The jaws have cross-cut face plates for better grip.

▲ A hobby bench vice

Metalworking vice
Metalworking vices are more commonly known as engineers' vices. They are designed in a similar way to woodworking vices, in that they are primarily used to hold pieces of material to ensure accuracy and safety when cutting material. Metalworking vices are designed to measure, mark and cut pieces of metal. Metalworking vices differ from woodworking vices in that they are raised from the workbench. Woodworking vices are flush to a workbench.

▲ A metalworking vice

Hand vice
Hand vices are designed to hold small objects tight while an engineer conducts the process of drilling, filing, hammering, sanding and shaping. The main purpose of the hand vice is to ensure the engineer's hands are safe when delivering the above processes.

Machine vice
Machine vices are designed to be employed when using machines. These vices can be bolted or clamped to the base of the machine when drilling or milling. They are clamped or bolted for safety and to ensure faster operations can be performed. For smaller work, they may be held down by hand.

▲ Hand vices hold small workpieces tightly.

G-clamp
A G-clamp, as its name suggests, is shaped like a letter 'G'. It is a versatile clamp as it can be used to hold many materials in position, such as plastic and metals. It can be bought in a range of sizes. G-clamps can also be used to secure machine vices to machines or benches.

F-clamp
Like the G-clamp, the F-clamp gains it name from its design as it looks like a letter 'F'. An F-clamp is suited to holding large workpieces, due to its wide opening capacity; it has one long, vertical bar and two horizontal jaws. F-clamps are suitable to use with both metal and wood.

▲ A machine vice

Trigger clamp
A trigger clamp is a versatile clamp that can be used to hold a range of workpieces and materials. A trigger clamp is designed for holding together parts of your workpiece while you are carrying out different tasks such as cutting, gluing, nailing and sanding. The trigger clamp can be used for a variety of different applications, from light-duty woodworking to heavy-duty manufacturing processes.

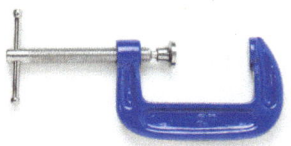

▲ G-clamps come in a range of sizes.

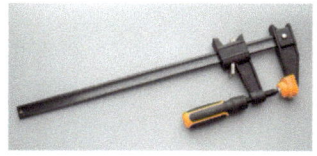

▲ F-clamps are used for securing large workpieces.

▲ A trigger clamp

▲ Sash clamps have holes for screws to fit into, making them easily adjustable.

Sash clamp/sash cramp

A sash clamp/sash cramp is a handy tool in woodworking, perfect for securing big pieces of wood. These clamps are designed to be long and adjustable, with holes for your screws to fit into, making them simple to resize and use.

Jigs

A jig is a tool that can be bought, but engineers often make their own jigs to enable an engineering process to be carried out more quickly, easily and safely. A jig can be created from a variety of materials such as wood, metal and plastic. It can make a process safer by ensuring that the material is held in place more securely, or it can remove the need for the material to be hand held. A jig can also ensure accuracy if it has pre-drilled holes to follow as a guide. Finally, it allows speed; a jig can remove the need to measure the workpiece – as this was completed when designing the jig – therefore making the process easy to repeat and carry out multiple times.

Cutting materials: hand held

Metals

Hacksaws

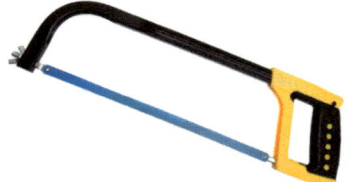

▲ Hacksaws have small, hard teeth making them very useful for cutting metals.

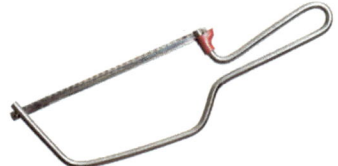

▲ Junior hacksaws are used in tight spaces.

Key Term

TPI Teeth per inch. The number of teeth a saw blade has per inch of length.

There are two main types of hacksaw: the standard hacksaw and the junior hacksaw. The junior hacksaw is a smaller version of the hacksaw and is sometimes easier to use on smaller jobs or in smaller spaces. (Plumbers sometimes use junior hacksaws to cut copper pipes because of the tight spaces they work in.) The hacksaw is mainly used to cut different types of metals but can also be used on certain plastics.

The teeth on a hacksaw are very small and hard, making them ideal for cutting harder materials such as metals. You can buy different blades with varying **TPI** that are used for cutting different types and thicknesses of metal (see table below). Most all-round jobs can be carried out with a 24 TPI hacksaw.

Teeth per inch (TPI)	Material/job
14 TPI	Thicker metals, softer metals
18 TPI	Thick to average metals
24 TPI	Average metals/general use
32 TPI	Thinner metals, harder metals

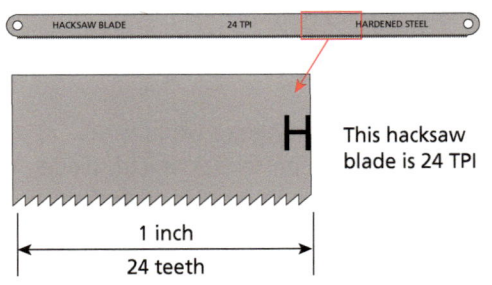

▲ Saw blades are measured in terms of teeth per inch.

Woods

Tenon saw

A tenon saw is a large **backsaw** used for cutting accurate lines in timber. (A backsaw has a metal spine or rib running along its back to stiffen it, enabling high precision in cutting.) Tenon saws are used for making straight and fast cuts. They are commonly used to make the tenons in **mortise and tenon joints**. These saws can be used on hard and soft woods. As standard, a tenon saw will have between 10 and 14 teeth per inch (TPI); this allows you to have more control over the depth and direction of the cut you are making.

▲ *A tenon saw, a type of backsaw*

Coping saw

A coping saw is used on timbers and plastics to cut intricate internal and external shapes. It is called a coping saw because of its use in coping or scribing the end of a length of moulding to make it fit against another.

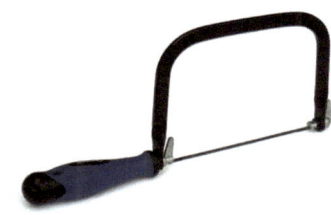

▲ *A coping saw*

Fretsaw

A fretsaw is a type of **bow saw**. It is similar to a coping saw as it is used to cut intricate work, which often incorporates tight curves. However, the fretsaw is capable of cutting much tighter curves and performing more delicate work.

Mitre saw

A mitre saw is designed for cutting timber and can be attached to a woodworking vice. The saw can be moved to different angles, therefore timber can be cut at various angles, creating a range of joints.

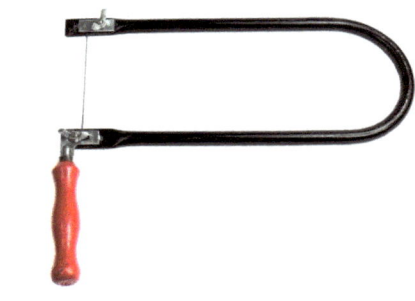

▲ *A fretsaw*

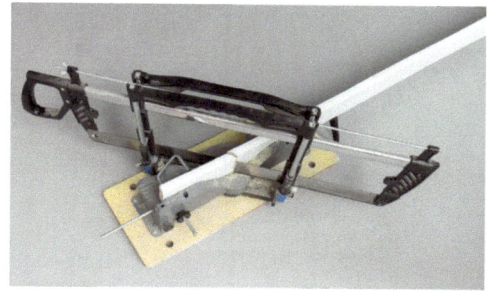

▲ *A mitre saw*

Filing materials

Hand files

Hand files are mainly used to smooth rough surfaces on metallic objects. They can be used on certain plastics (although some plastics clog up the file) but should not be used on woods; there are other tools, such as rasps, that perform the same job for woods. Just like sandpaper, hand files use an abrasive surface to smooth the material and come in different grades (e.g. rough or smooth). When working on a workpiece, you should start with a 'rough'-grade file and eventually finish with a 'smooth'-grade file. Files are used for shaping and finishing metals.

There are many differently profiled/shaped files that are used for various jobs.

▲ Detail from a rough-grade file

▲ Some of the different shapes of file that are available: round, square, triangular, half-round and flat files

Flat files

The flat file is the most common file and is used to smooth workpieces flat. It has a 'safe edge' that is smooth so only one surface of an internal corner can be filed.

Semi-flat/half-round files

The semi-flat or half-round file is used on interior curves.

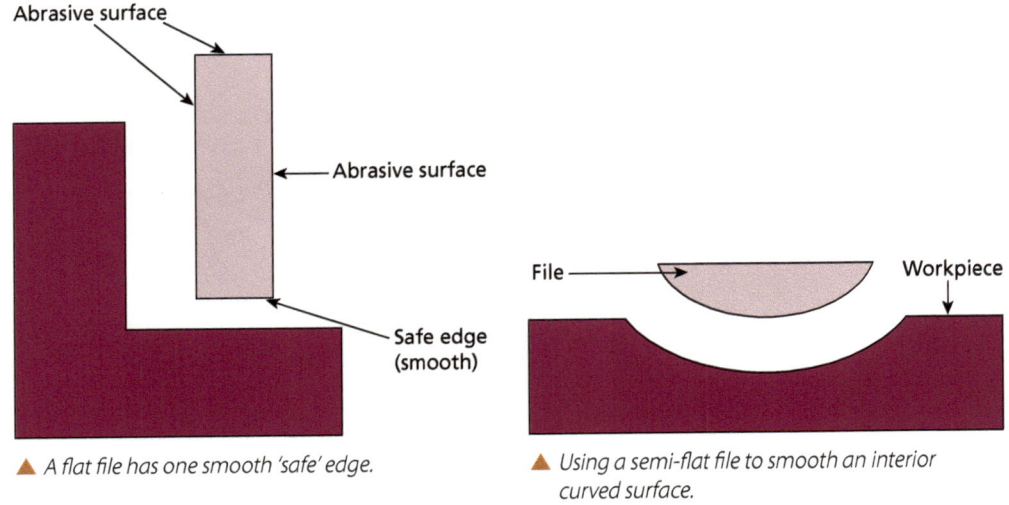

▲ A flat file has one smooth 'safe' edge.

▲ Using a semi-flat file to smooth an interior curved surface.

Round files

The round file is used on the interior of drilled holes. It is good for removing burrs.

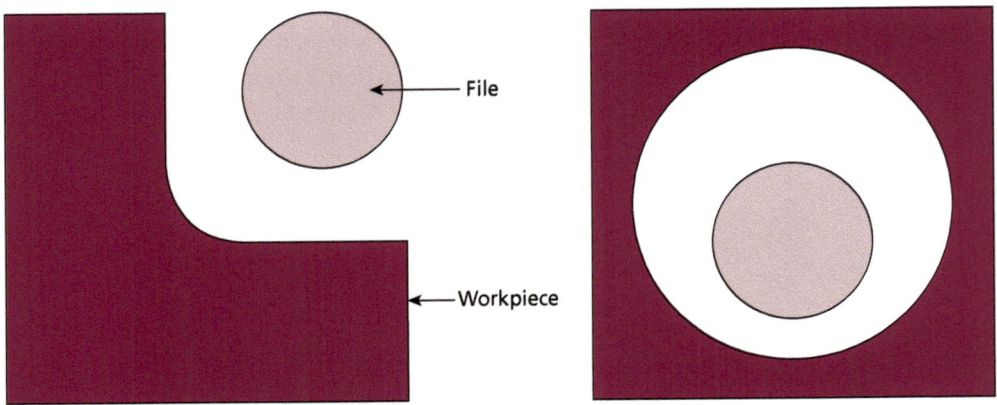

▲ A round file smooths the interior surface of drilled holes.

Triangular files
A triangular file can fit into very tight internal corners. It is also useful for starting a 'groove' on a flat surface.

Square files
A square file is good for filing internal corners (both edges). It can also be used to file square grooves into the surface, as it is quite thin in profile.

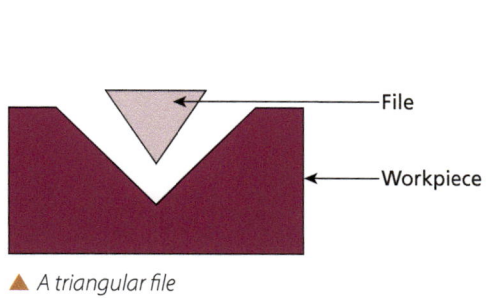
▲ A triangular file

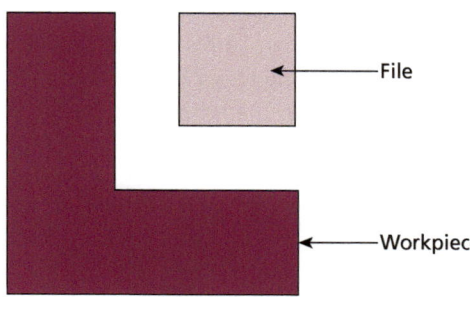
▲ A square file is useful for smoothing internal corners.

Needle files
Needle files are small files that are used to finish and shape metal. They come in the same shapes as the hand files above and can be fine, medium, coarse and extra coarse. One side of the file is smooth so that when filing in tight spaces, it does not mark metal that you do not wish to file.

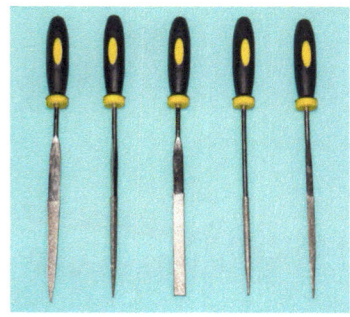
▲ Needle files are used to finish and shape metal.

Rasps
A rasp is a form of file that is used for shaping wood. It is typically a hand tool that is in a coarse form. Rasps often consist of a tapered rectangular, round or half-round sectioned bar with individually cut teeth.

▲ A rasp

Files vs rasps
The main purpose of files is to smooth and finish metals and certain plastics, while rasps are used on timber. Rasps can make deeper and coarser cuts than files because, unlike files, they have separate and triangular, rather than parallel, lines of teeth. Rasps remove the wood, as opposed to planing it away as files do. Files have spaces between the teeth that, if used with wood, would quickly fill with wood chips; in contrast, a rasp is well suited to shaping wood quickly.

Shaping materials

Ball-pein hammer
A ball-pein hammer can be used to shape metal. It is also used for traditional processes, such as striking a centre punch, by using the flat surface. The rounded part of the head can be used to shape sheet metals or to shape the heads of rivets. Most shaping is now completed with industrial machines but the ball-pein hammer is good for small jobs in a workshop environment. Ball-pein hammers have to be tough, so are made from forged high-carbon steel (heat treated).

▲ A ball-pein hammer is used to shape metal.

Planishing hammer
A planishing hammer is used for smoothing metal. The flat side will refine the hammer marks from the rounded side of the hammer to a nearly smooth finish. This hammer is also used to size rings, form bezels and for general forming of metal against mandrels and stakes (see page 81).

> **Key Term**
>
> **Planishing** A metalworking technique that involves finishing the surface of sheet metal by finely shaping and smoothing it.

▲ A doming block and punch set

▲ A rubber mallet

▲ Tin snips

▲ Pliers

Doming block and punch set
A doming block has a range of different-sized indentations that you can choose from depending on the design you have in mind. Doming punches are available in a range of sizes and the most suitable sized punch is used alongside a hammer to strike the metal and start doming the piece.

Anvil
An anvil is used to shape metal. A piece of metal is placed on the anvil and shaped by hand with a hammer. A blacksmith's anvil is usually made of wrought iron or cast iron, with a smooth working surface of hardened steel. A projecting conical beak, or horn, at one end is used for hammering curved pieces of metal.

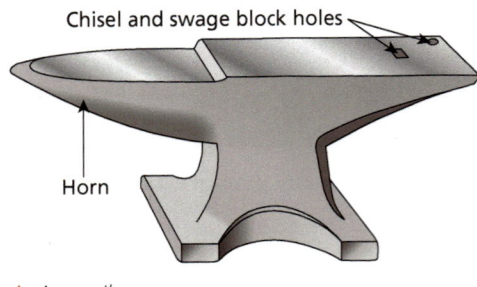

▲ An anvil

Mallet
A mallet is a block on a handle, which is usually used for driving chisels. On a rubber mallet, the head or block is made of rubber. These types of hammers deliver a softer impact than a hammer with a metal head. They are essential if your work needs to be free of impact marks.

Tin snips/hand shears
Tin snips are used to cut sheet metals by hand. There are larger guillotine-style shears for larger gauge (thicker) sheet metal, but these are cumbersome and generally fixed in one place. Tin snips are easy and quick to use by comparison but there is a limit to the thickness of metal that they are able to cut.

Pliers
Pliers come in a range of shapes and sizes. They are designed for many uses, including gripping something cylindrical like a pipe or rod. They are useful when twisting and loosening fixtures on materials and can also be used for twisting and cutting wires.

▲ A tapered reamer widens holes, leaving smooth sides.

Tapered reamer
A reamer tool is designed to widen a pre-existing hole in a piece of metal that is too small. The material will only be slightly widened and the reamer will leave smooth sides by removing rough edges or burrs. Reamer tools are hard and durable.

Deburring tool
A deburring tool is designed to remove burrs and jagged edges on the surface of a piece of material or an object that has been created during the machining or casting process. A deburring tool allows the engineer to create a product with improved quality and functionality. It can be used on wood or metal products.

▲ A deburring tool can be used on wood or metal workpieces.

Sheet-metal benders

Metal benders are designed to make accurate bends in sheet metal. This is essential to create clean edges of metal. The sheet metal is placed between the two rollers that roll back and forth while applying pressure.

Threads

Tap and die

Tap and die sets are used to create threads on or in workpieces. With a tap and die set, you can create nuts and bolts from different types of metals. Tap and die sets can also be used to clean up existing threads by a process known as **chasing** but this tends to leave the thread a little looser.

Tap

A tap is used to cut or create **internal** threads in a drilled hole. When a hole has been drilled, you can use the tap to cut a thread on the walls of the hole by effectively screwing the tap into the drilled hole. Taps use a **tap wrench** that is adjustable for different sizes and there are different-shaped taps available. Tapered taps are used to make it easier to start the cutting action.

▲ A sheet-metal bender makes accurate bends in sheet metal.

▲ Tap wrenches are used for creating internal threads.

▲ An internal thread

Die

▲ A die, die wrench and taps. Dies are used to create external threads.

▲ An external thread

A die is used to cut or create **external** threads. You can fit the die over the end of a piece of round-section metal and, with a rotational motion, slowly cut a thread on the exterior surface of the metal section. Dies can be slightly opened or closed by using the screws on the die holder/wrench. This helps with starting the screwing and allows you to adjust the diameter setting to fit a piece of round-section metal. To create an M8 thread, you would typically need to drill a 7 mm hole to allow enough material for the thread to be cut. Note that drilled hole sizes differ depending on what size thread you need to cut, so it is always advisable to check a chart. The following table offers a rough guide to help you.

Tap size	M3	M4	M5	M6	M7	M8
Drill size	2.5 mm	3.3 mm	4.2 mm	5 mm	6 mm	6.75 mm

Key Terms

Chasing The act of re-cutting a thread with a tap or die to repair any damage such as cross-threading. The process can also be used to clean up the thread if it is old, worn or dirty.

Tap wrench A device for holding a tap. A tap wrench has arms with a textured surface to improve grip and can be rotated by hand.

Top tip

You will often see information on the taps and dies to explain the sizing. For example, M8 = metric 8 mm.

Cutting compound

When using tap and die sets to create internal or external threads, it is always advisable to use cutting compound on your tools. Cutting compound ensures that you achieve a smooth clean cut edge, while also making the cutting operation easier. Cutting compound will also prolong the life of your tools.

Assembling materials

Screwdrivers

▲ Types of screwdriver showing the different types of tip

Screwdrivers are hand-held manual tools that come with a number of different tips. These tips allow a variation of screw heads to be submerged into different materials by applying pressure and turning. Screwdrivers can be bought on their own, as part of a set of screwdrivers or with interchangeable tips which can be changed depending on the screw that needs to be fastened.

Hex keys

▲ Hex keys or Allen keys

A hex key, more commonly known as an Allen key, is a driver for bolts or screws that have heads with internal hexagonal recesses. A hex key is a hand tool that is used for tightening and loosening hexagonal bolts.

Spanner

A spanner is a metal hand tool and is a type of wrench. Both ends of the spanner fit round a nut so that you can turn it to loosen or tighten it.

Adjustable wrench

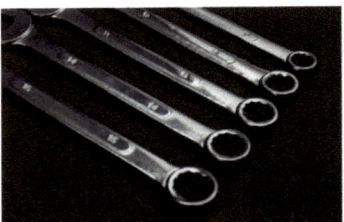
▲ Spanners

The purpose of a wrench is similar to that of a spanner. Whereas spanners come in a set size that can be used on a specific sized nut or bolt, an adjustable wrench can be adjusted to fit many different sized nuts and bolts.

Ratchets and sockets

A ratchet and socket is more commonly known as a socket wrench. It is similar to a wrench and spanner in that it allows you to turn a nut or bolt. However, the key difference between a socket wrench and a spanner or wrench is that you can turn the nut or bolt without repositioning the tool on the fastener. This allows you to turn nuts or bolts in tight spaces, as the socket wrench does not need to turn full circle. The sockets can be changed to ensure a large range of nuts or bolts can be fastened or loosened. Some ratchet wrenches can also have torque settings, allowing you to tighten a bolt to a specific torque setting. These are known as 'torque wrenches'.

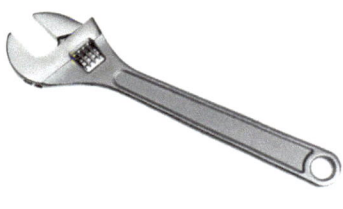
▲ A wrench

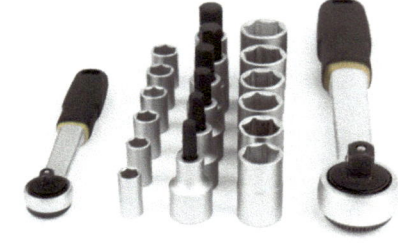
▲ Ratchets and sockets

Task 3.1

In your notebook, identify the different tools and equipment illustrated, and their use/function.

Engineering machines

The centre lathe

The centre lathe is a machine used to manufacture mainly cylindrical products or objects. It can use different 'sections' of metal (e.g. a hexagonal section, square section, etc.) to produce shapes such as cubes, but these operations tend to be more technically advanced than just creating a cylindrical object. Centre lathes are operated both manually (in workshops) and by way of computer numerical control (CNC) in industry. Think about how many items/parts there are in the world that are cylindrical. Many different materials can be used on a centre lathe such as metals and plastics. The figure below is an image of a centre lathe that you would typically find in a manual workshop.

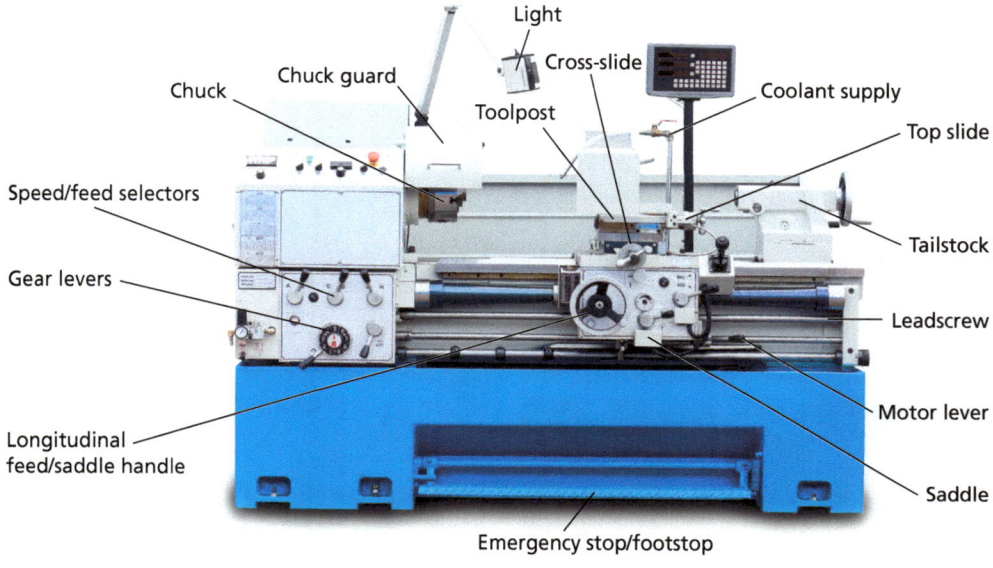

▲ A centre lathe

Centre lathe processes

Facing off

Facing off is the most basic operation of the centre lathe. A piece of metal is placed into the three-jaw **chuck**. The chuck is the part of the centre lathe used to hold the workpiece and drill bit in the **tailstock**. The lathe cutting tool is shaping the material's surface and this can also be assessed for precision. You can use the measurements on the handles of the centre lathe to identify how much material has been removed.

Turning

Turning, or turning down, is the most common of the centre lathe operations. In this process, the cutting tool will be applied to a cylindrical object and it will remove material until the diameter is the desired size.

Parting

Parting, or parting off, is the stage you complete when your part has been made. The parting tool is applied to the completed part, slowly, until the completed part has been cut away. This gives a clean-cut finish to your part. The rest of the material is left in the three-jaw chuck.

3 Using engineering tools and equipment

 Top tip

When first starting out on the centre lathe, do not attempt to use multiple handles at once. Use both hands on one handle for precision.

 *Key Term*

Tailstock The part of the centre lathe that sits towards the rear of the machine and holds various useful tools such as chucks, drill bits and centre guides. It can also support longer workpieces to stop them from wobbling when they are being rotated.

▲ *Facing off on a lathe*

▲ *Turning, or turning down, is a common operation on a centre lathe.*

▲ *Parting, or parting off, is carried out when your part is complete.*

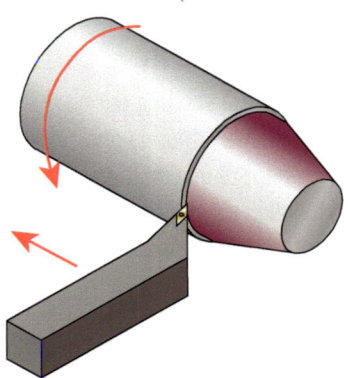

▲ Taper turning creates a taper down the length of a workpiece.

Taper turning
Taper turning is the process of creating a taper down the length of the workpiece. To create this desired outcome, the cutting tool moves at an angle to the workpiece. In a tapered piece, the diameter of the workpiece changes uniformly from one end to another.

Knurling
Knurling uses a tool which is quite different from other tools commonly used on the centre lathe. The knurling tool is applied to the rotating workpiece and the pressure places a criss-cross textured pattern onto the surface of the workpiece. The textured patten allows for extra grip and is useful for items such as handles.

▲ Knurling on a centre lathe to create a criss-cross pattern

Drilling
A centre lathe can also be used to drill centred holes into workpieces accurately. It works in a different way to a conventional drill as the workpiece rotates while the drill bit remains stationary. The drill bit is held in a standard Jacobs chuck that is, in turn, held into the spindle of the tailstock.

Boring
Boring is the process of enlarging a hole that has already been drilled. Boring is a cutting operation that uses a single-point cutting tool or boring head.

▲ Drilling workpieces using a centre lathe

Grooving
Grooving is an operation that creates a narrow cut or groove into the workpiece. The size of the cut depends on the size of the cutting tool selected. There are two types of grooving operations: external and face grooving.

▲ Boring

▲ Grooving

Thread cutting
Internal and external threads can also be created on the centre lathe, but specific thread-cutting tools are necessary for this operation. The process of creating threads is automated and should be carried out using the automatic feed system on the centre lathe. Alternatively, to create an external thread, you could place the workpiece in the three-jaw chuck, as it self centres, and use a die manually to create the thread. You could also create an internal thread this way, by applying a tap to the tailstock and twisting this manually inside the workpiece, which is again attached to the three-jaw chuck.

▲ Thread cutting: external

▲ Thread cutting: internal

3 Using engineering tools and equipment

The vertical miller

A manual milling machine, the vertical miller is used to shape materials such as metals and plastics. Most products manufactured with milling today use CNC millers to shape different materials (such as aluminium phone carcasses). Milling can be very accurate when done correctly as the cutting tools are changeable. You can also mill using small-diameter cutting tools for greater levels of detail.

▲ Milling

▲ Milled aluminium

▲ A vertical miller, used to shape metals and plastics

Milling machine processes

Face milling
The material is placed into a machine vice and the tip of the cutting tool is used to cut the surface of the workpiece. The size of the cutting tool will be dependent on the size of the workpiece. Best practice is to ensure the cutting tool is used in sections so excess **swarf** is removed from the tool. The outcome is an accurate and flat surface which has been cut to the correct thickness.

Peripheral milling
Peripheral milling differs from face milling in that it uses the side of the cutting tool to cut the material, as opposed to the tip. In peripheral milling, the cutting tool is placed parallel to the material. The tool is positioned so that the side of the cutting tool cuts away at the edges or faces of the material.

> **Top tip**
>
> The milling machine can go in various directions:
> - X: side to side (left to right, right to left)
> - Y: back and forth (forward to back, back to forward)
> - Z: up and down, down and up.

> **Key Term**
>
> **Swarf** Small pieces of metal or debris produced as a result of the machining process.

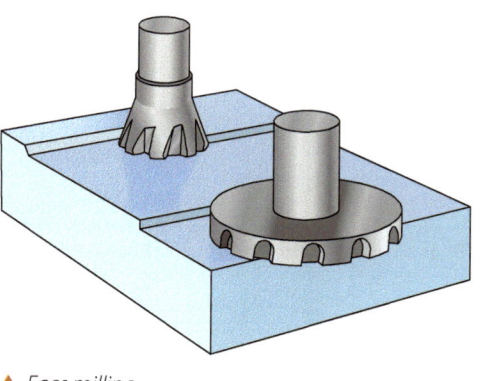
▲ Face milling

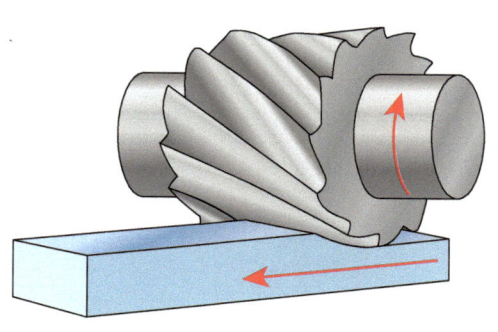
▲ Peripheral milling

Slot milling
Slot milling, also known as groove milling, is when the cutting tool makes a slot or groove within the piece of material. Slots can be short or long, closed or open, narrow or wide. These parameters will be determined by the size of the cutting tool.

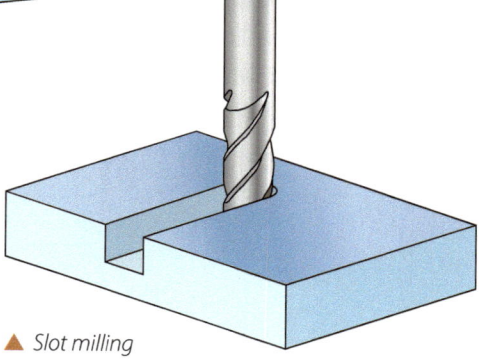
▲ Slot milling

55

End milling

End milling uses a specific cutting tool consisting of a cylindrical cutter that has multiple cutting edges on the periphery and the tip. This allows the cutter to end cut and peripheral cut. These cutting edges or flutes are usually made helical to reduce the impact that occurs when each flute engages the workpiece.

Machine drills

Machine drills are drills that are fixed in one place. Unlike hand drills (such as cordless drills), machine drills can be very accurate as the workpiece can be clamped down or held in a machine vice while the rotating drill bit is lowered using the feed lever.

Machine drills include:
- the bench drill, a smaller type of drill that is bolted to a desk or bench
- the pillar drill, a larger drill that stands on the workshop floor.
  The pillar drill is more powerful and can therefore be used to drill larger-diameter holes.

All machine drills have a changeable belt system that allows the user to speed up or slow down the speed of the drill bit, depending on what material is being drilled and the diameter of drill bit being used.

Feed and speed rates for machines

Cutting speed is considered to be the speed of a tool that cuts the workpiece. In comparison, the **feed rate** is the distance travelled by the tool in one revolution. Therefore, the feed rate is considered as the velocity (speed) at which the cutter is fed.

Engineers can use sheets like the ones shown in the tables that follow to identify the correct speeds to use when cutting a variety of materials at different sizes. This is essential when using engineering machines such as pillar drills, centre lathes and milling machines, as this will ensure safe practice, accuracy and a good quality finish.

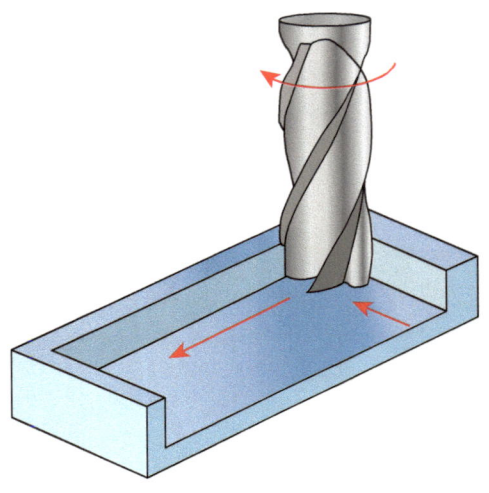

▲ End milling

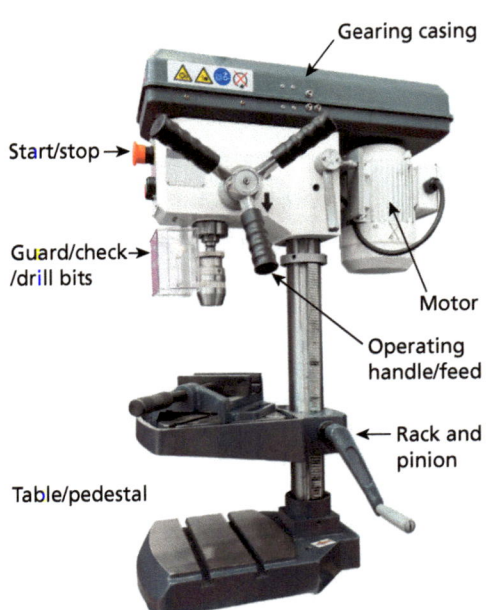

▲ A bench drill is bolted to a desk or workbench.

 **Top tip**

Speed and feed rate information should be placed by the different machines. This will allow the user to identify the correct speed for the material and size on the machine.

Guideline RPM for centre lathe					
Material diameter		Material			
Inches	Millimetres	Aluminium	Brass	Mild steel	Stainless steel
$\frac{1}{2}$	12.5	1400	1200	1000	600
1	25.5	700	600	500	300
$1\frac{1}{2}$	38	500	400	300	200
2	50.5	350	300	250	150
$2\frac{1}{2}$	63.5	280	250	220	120
3	76	225	190	160	100

3 Using engineering tools and equipment

Guideline RPM for milling machine					
Cutter diameter		Material			
Inches	Millimetres	Aluminium	Brass	Mild steel	Stainless steel
1/4	6.35	2500	1200	450	375
3/8	9.5	2200	600	375	300
1/2	12.7	2000	500	300	200
5/8	15.8	1750	400	250	150
3/4	19	1500	300	220	120
1	25.5	1000	200	150	100

Guideline RPM for pillar drill					
Drill bit diameter		Material			
Inches	Millimetres	Aluminium	Brass	Mild steel	Stainless steel
1/4	6.35	1200	1200	750	500
3/8	9.5	900	600	500	375
1/2	12.7	600	500	250	250
5/8	15.8	360	400	200	150
3/4	19	300	300	150	120
1	25.4	180	200	100	80

 **Top tip**

Knurling and parting operations should only be run on a centre lathe at 100 RPM maximum.

 Top tip

Counterbore bits and operations should run at 200 RPM maximum, while countersink bits and operations should be run at 300 RPM maximum.

 **Key Term**

LED Light-emitting diode; an electrical component that gives off or emits light.

Hand drills

As the name suggests, hand drills are drills that are held by hand. There are many types of hand drills available (microdrills, brace drills, egg-beater style, etc.) but one of the most common is the corded/cordless hammer drill. These drills can be purchased from any DIY store and most households have a version of this type of drill. They are used mainly to drill into masonry (walls) to hang shelves and curtain rails. These drills come with a hammer setting that moves the drill bit in and out while it rotates, to create a chiselling action. This makes it easier to drill through masonry. They also have torque settings and can have many different features such as keyless chucks (see the following section), **LED** lights, magnets, variable power/speed settings and front handles for big holes (mainly SDS drills). Cordless drills are very popular now as they can be just as powerful as corded drills but without the problem of having to set up extension cables wherever you are working.

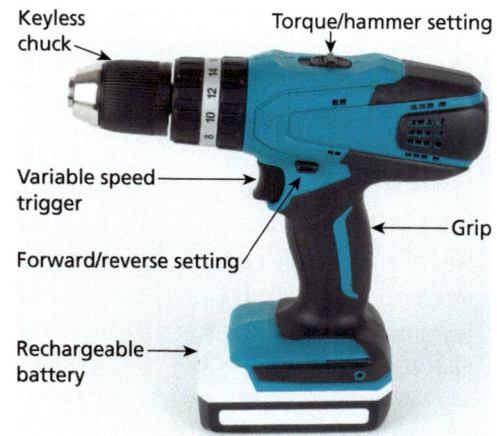

▲ *A cordless hammer drill*

 **Top tip**

Depending on what drill bit you are using, you can use a hand drill to drill into most resistant materials.

▲ *Drill bits*

▲ *Three-jaw centre lathe chuck*

▲ *Four-jaw centre lathe chuck*

Chucks

A chuck is the part of a drill or machine that holds the drill bit (cutting tool). In general, corded drills (as well as machine drills) have chucks that need to be used with a **chuck key**. The chuck key is used to loosen/tighten the jaws of the chuck and can be tightened up to a high torque setting for bigger diameter drill bits.

On cordless drills you will mainly see **keyless chucks** that can be tightened by hand. These types of chucks rely on a strong grip from the user to ensure a high torque fit for a drill bit. The cordless chucks have the added advantage of not having an extra part/component to carry or use and there is no danger of not being able to use the drill because of a lost chuck key.

Most chucks are three-jaw ones that are **self-centring** when using round or hexagonal sections. However, chucks on centre lathes can be changed to a four-jaw chuck for square/octagonal sections. These four-jaw chucks need to be centred manually.

The Jacobs chuck, invented by Arthur Jacobs and patented in 1902, is a term often used for the most commonly used chuck.

> **Key Term**
>
> **HSS** High-speed steel. A type of steel that is very hard and contains a higher percentage of carbon than standard high-carbon steels. It has added elements like tungsten, molybdenum and vanadium. It is often used for cutting tools such as drill bits and saw blades. While HSS is sometimes referred to as a high-carbon steel, due to its carbon content, it is a specific and distinct type of steel known for its excellent cutting and tool performance, especially at high speeds.

Drill bits

Drill bits are the cutting tools that are placed into the chuck of a drill and rotate to cut a hole in a resistant material such as wood, metal, plastic or masonry. There are many different types of drill bits available that can be used for a variety of different jobs. Below are some common drill bits that you might find at home or in a workshop.

- **Twist drill bits** are one of the most common drill bits around. They can drill most materials including metals, plastics and wood (although there are other specialist wood bits) but are no use when drilling into masonry. They are made from **HSS** which is more resistant to heat than other high-carbon steels and therefore does not wear as much. Twist drill bits are often dark grey/black in colour.
- **Masonry drill bits** are very common in households and are used for drilling into bricks/walls/masonry. They have a **chiselled** tip, often made from tungsten carbide (a very hard material), that is joined to the steel shaft. The chiselled tip helps to chisel away the masonry when using the hammer setting on the drill. Masonry bits are often shiny silver in colour.
- **Flat drill bits** are normally found in woodworking workshops and are used to drill larger-diameter holes in wood boards and sections. They leave rough edges and should only be used with a higher-power drill due to the friction involved. They are not to be used on metals.
- **Forstner drill bits** are generally used to cut **blind holes** (where you do not drill all the way through the wood) that are useful for hinges on different types of furniture with doors (e.g. kitchen cupboards). They come in larger diameters and should only be used with timbers (man-made and natural) or some plastics.

▲ *Twist drill bits*

▲ *Masonry drill bit*

▲ *Flat drill bit*

▲ *Forstner drill bit*

Buffing/polishing machines

A buffing machine (sometimes known as a polishing machine) is used to put a final polish or finish on a workpiece. It is mainly used to bring metals up to a highly polished finish but it can also be used on some plastics. Before using this machine, your metal workpiece should first have any scratches or vice marks removed with fine files and **wet and dry paper** to present a smooth surface for the buffing machine. The polishing parts of the machine are called **mops**. These are natural fabric discs that are stitched together with many layers. When they are spun at high speeds (up to 3000 RPM) they present a rigid surface that can be effective at polishing metals when they are pressed against them.

▲ A buffing or polishing machine can be used to put a finish on a workpiece.

Wet and dry paper

Wet and dry paper comprises abrasive sheets of paper (similar to sandpaper) for use mainly with metallic surfaces. As the name would suggest, it can be used either wet or dry. When wetted, the moisture acts as a lubricant and removes the particles more quickly than when dry, creating a smoother surface faster. Like abrasive tools and equipment (files, grinding wheels, sandpaper), wet and dry paper comes in different grades (grit sizes) for a rougher or smoother finish. The lower the number/grit size of the wet and dry paper (e.g. 40 grit), the coarser the particles are on the paper and the rougher the finish; the higher the grit size (e.g. 1000 grit), the finer the particles are on the paper and the smoother the finish.

 Top tip

Polish should be applied to the mops to ensure a better finish.

Grinders

A bench grinder can be found on a worktop bench. It has abrasive wheels and is used to hand grind and sharpen a range of cutting tools, such as tool bits, drill bits, chisels and gouges. A secondary function of a bench grinder is to provide a rough shape for metal prior to cutting or welding. You can also find a grinding machine in a larger version that is mounted to a pedestal.

▲ A pedestal grinder

 Top tip

For any piece of large machinery, please ensure that all health and safety rules are followed: hair should be tied back, and goggles and appropriate PPE should be worn (see page 66).

Angle grinder

An angle grinder can also be called a side grinder or a disc grinder. An angle grinder is a hand-held power tool used for cutting, grinding, deburring, finishing and polishing.

▲ A bench grinder   ▲ An angle grinder

Metal bandsaw

A metal bandsaw is a powered saw that has a moving blade with serrated toothed edges. The main purpose of a metal bandsaw is to make accurate cuts in metal. It is ideal for cutting all types of metals – **ferrous** and non-ferrous – with varying thicknesses.

▲ A metal bandsaw can cut all types of metals.

Key Term

Ferrous A ferrous metal is any metal that contains iron.

 **Top tip**

Grinders are extremely dangerous and should not be used by engineering students in school.

Guillotine

Sheet-metal guillotines are used to cut sheet metal into a specified size, where precision is required. The thickness of the metal can vary and the sheets can be large. The guillotine uses a blade to cut through sheet metal by applying pressure on the handle and therefore both sides of the blade. This makes it very efficient at cutting through thick pieces of steel or aluminium quickly.

Woodworking tools and machines

Bandsaw

Bandsaws are commonly found in workshops, in industry and in schools. They are primarily used for woodworking, but by switching to a blade with more teeth (TPI), they can also cut metal (see above). However, when available, it's recommended to use a dedicated metal bandsaw, specifically designed for accurate and efficient metal cutting. Like the wood bandsaw, it's well-suited for cutting metal in large quantities with precision.

▲ A guillotine cuts sheet metal into specified sizes.

Tablesaw

Tablesaws are used for cutting large boards or pieces of timber into straight lines or smoother cuts. A tablesaw is a common piece of equipment in industry and school workshops. Projects that are cut on the tablesaw normally require a more precise cut than those crafted using hand-held tools. A tablesaw can be adjusted depending on the angle or size of cut that you require.

> **Key Term**
>
> **Linisher** A machine that improves the flatness of a surface by sanding or polishing it. It is also known as a linish grinder.

▲ A bandsaw is primarily used for cutting wood.

▲ A tablesaw is commonly found in workshops.

Scroll saw

Scroll saws are used for cutting intricate curves or corners in timber. They are often used to complete a project due to their speed and accuracy. The scroll saw is a stationary machine which is normally attached to the work bench; the timber is, therefore, moved around the blade.

Linishers

Linishers, also known as sanders, can come in various sizes and shapes. The more common linishers you would see in a school workshop would be a belt sander or a disc sander. Linishers turn an abrasive sheet across the material, smoothing and levelling the surface. They are mainly used with wood, but can also be used for plastics. Linishers can often be used to prepare timber before a finish is applied.

▲ A scroll saw can cut accurate curves and corners in wood.

3 Using engineering tools and equipment

▲ A belt linisher or sander can smooth and level a timber surface before finishing.

▲ A linisher disc

Router

Routers are power tools commonly used in woodworking. They can be hand held or mounted to router tables. A router has a flat base with a rotating blade extending past the base. This blade routs – hollows out – an area in a material such as wood or, less commonly, plastic. It is an extremely versatile piece of machinery.

Planers

A planer can be hand held or fixed to the floor. Hand-held options can come in manual or electric form. A planer is a common woodworking machine used to trim boards to a consistent thickness throughout their length.

▲ A router tool

▲ A manual planer

▲ An electric planer

> **Top tip**
>
> All large, electric woodworking machinery will require extraction to ensure the majority of dust particles are collected.

Chisels

Chisels can come in varying sizes. A chisel is a cutting tool with a sharpened edge at the end of the metal. The purpose of a chisel is for carving and cutting hard materials such as wood and stone. A sharp wood chisel can cut mortises, shave rough surfaces, chop out corners and scrape off glue.

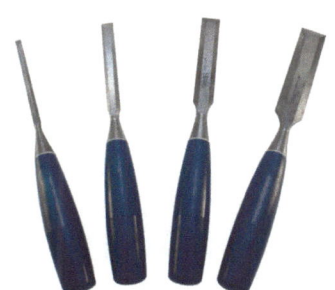

▲ Chisels are used to carve and cut hard materials.

Wood-turning lathe

Wood-turning lathes, as the name suggests, are mainly used to shape wood into cylindrical profiles. Common projects made on a wood lathe are items such as furniture legs, baseball bats and bowls. Wood-lathe tooling consists of devices to fix and secure the workpiece, a moveable tool rest, and hand-held cutting tools in the form of long-handled gouges, skews, scrapers and parting tools.

▲ A wood-turning lathe

61

Etching

UV lightboxes and PCB etching tanks

A number of workshops will have the equipment needed to manufacture their own printed circuit boards (**PCBs**). With the correct knowledge, most engineers can design their own circuits to perform simple functions. To create the simple circuits, engineers will have to go through the process of making a circuit board. To make a simple circuit board you will need to do the following:

1. Design a simple circuit.
2. Cut some copper-plated photoresist board to size.
3. Place a printed image of your circuit onto a thin plastic sheet (masking the circuit area on the copper plating).
4. Expose the photoresist board to ultraviolet (UV) light (using a UV lightbox).
5. Place your photoresist board into developing solution.
6. Then place your photoresist board into a PCB etching tank (filled with etching fluid) where the unwanted copper will dissolve away leaving only the copper tracks for the circuit you designed.
7. Start to populate your PCB with the needed components.

> **Key Term**
>
> **PCB** A printed circuit board. This type of circuit board is printed using computer-aided manufacture.

Task 3.2

In your notebook, identify the engineering machines and their use/function.

CAD/CAM

CAD (computer-aided design) is the use of computers (or workstations) to aid in the creation, modification, analysis or optimisation of a design.

Computer-aided design (CAD)

Uses of CAD

The way in which CAD can be used depends on the type of engineering it is being used to support. The lists below detail the way it can be used in structural engineering and design engineering.

Structural engineering
- Different elevations – views, 3D, floor layouts
- Different finishes/renderings
- Editing and changing details
- Use of mathematical calculations for structure
- Virtual reality views/tours of rooms

> **Key Terms**
>
> **CAD (computer-aided design)** Computer software for designing products in readiness for CAM.
>
> **CAM (computer-aided manufacturing)** Machines manufacturing parts and products using a computer program.

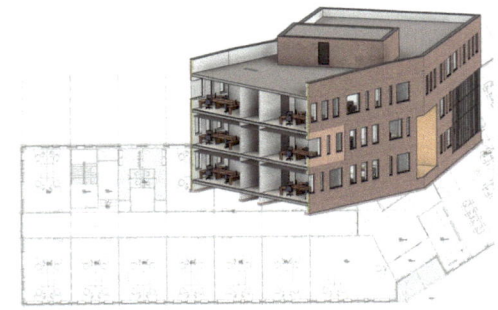

▲ Features of CAD – structural engineering

Design engineering
- Draw 3D views of products
- Change materials/colours of a CAD model
- Presentation of ideas/products to clients – CAD enables clients to see how something would look in real life
- Ability to download designs to a **CAM** (computer-aided manufacturing) machine and so create the product out of material

Advantages of CAD
- Files are easily stored and can be transferred electronically via e-mail.
- Drawings can be changed easily, removing the need to redraw.
- Scaling of drawings is possible.
- Drawings can be more accurate than those produced by hand.
- Drawings can be accessed anywhere.
- Reduced physical storage space is required to house drawings.
- Drawings can be viewed from all angles.
- Can create stress calculations to test designs, such as bridges.
- Can easily display drawings in presentations and on large screens.
- Colour hatching or rendering can be used to see how products would look in real life.

▲ *Uses of CAD – design engineering*

▲ *Advantages of CAD*

Advantages of CAD to engineer
- Quicker progression of design development, as minor changes can be made on screen.
- The design can be viewed from all angles.
- Finishes can be changed to make design more aesthetically pleasing.

Advantages of CAD to client
- Product can be viewed with different finishes and in different environments.
- Product can be viewed as a graphic file attachment to an e-mail.
- Client can make decisions about the product before manufacturing starts.

Advantages of CAD to manufacturer
- Ability to see what processes are required to make the product.
- Manufacturer can work out which material is the most suitable to use.
- Ability to plan ahead in terms of machinery or equipment that will need to be purchased.
- Ability to plan the jigs or components needed to complete the product.

Disadvantages of CAD
- Work can be lost because of the sudden breakdown of computers.
- Work is prone to viruses or could be easily 'hacked'.
- It takes time to learn how to operate or run the software.
- High production or purchasing cost for new systems.
- Costs of training the staff who will work on it.
- Regular updates are needed for software or operating systems.
- CAD/CAM systems may result in fewer staff being required.

Computer-aided manufacturing (CAM)

Also known as computer-aided modelling or computer-aided machining, CAM is the use of software to control machine tools in the manufacture of workpieces from computer designs. Drawings are downloaded from CAD software to machines that are connected to the computer. A common example of CAM in schools is the laser cutter. This machine cutter uses drawings made in 2D design (CAD) to cut shapes out from acrylic.

Relationship between CAD and CAM

CAD designs an object using a computer and CAM makes the object using a computer-guided machine such as a laser cutter, milling machine, lathe or drill. These machines are controlled by a computer.

Features of CAM

- CAM enables you to create a 3D product you can handle from a CAD drawing on a computer.
- Input from CAD packages is used to control a machine to make products.
- CAM is used to programme/control machines and robots including drills, mills and cutting machines.
- Products are made automatically 24/7.

Advantages of CAM

- Items are manufactured to a high degree of accuracy.
- Prototypes can be created quickly.
- The design modelling process is quicker, enabling the product to reach the market quicker.
- Lower costs for the manufacturer once the machines have been purchased, as fewer staff are required.
- Cheaper and better quality products for customers.

Disadvantages of CAM

- Software is very expensive.
- CAM equipment is very expensive to purchase.
- CAM equipment can be expensive to repair.
- Training will be required to use the equipment.
- Job losses will be necessary as the machines can work 24/7 and so reduce the need for staff.

CAM machines

3D printer

3D printing is also known as **additive manufacturing**. The process of 3D printing starts with a digital file that has been designed by an engineer. This is then printed on the 3D printer to a create a solid 3D object. To create the 3D object, layers of material are laid down by the printer until the object is completed. They are ideal for creating quick prototypes, but high-end printers can also be used for final projects.

Laser cutter

A laser cutter is a machine commonly found in school engineering workshops. A laser cutter uses a type of thermal separation process where a laser beam hits the surface of the material, which could be a range of plastics or woods, so strongly that it melts or cuts the desired outcome. An engineer will create their design on a digital file, before submitting this to the laser cutter to be cut.

CNC: computerised numerical control

CNC is a computerised manufacturing process in which pre-programmed software and code controls the movement of production equipment, such as centre lathes and milling machines. We have looked at these machines in a manual format, but now we will look at the CNC versions.

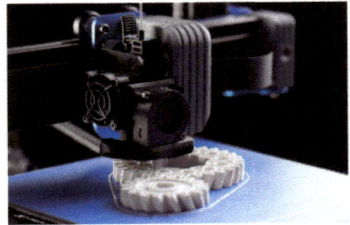

▲ A 3D printer builds up a 3D object by laying down layers of material.

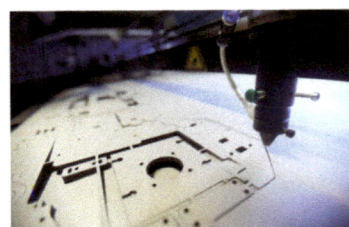

▲ A laser cutter melts or cuts a workpiece using a strong laser beam.

▲ CNC: computerised numerical control

CNC router

A CNC router is a machine that uses computer programming to control a high-speed rotating cutter to perform cutting and shaping operations. It can produce many types of product, such as moulds and formers for forming processes.

CNC milling machine

CNC milling machines are machine-operated cutting tools that are programmed and managed by CNC systems to remove materials accurately from a workpiece. The end result of the machining process is a specific part or product that is created using CAD software.

▲ A CNC router performs cutting and shaping operations.

▲ A CNC milling machine

CNC centre lathe

CNC lathe machines, or CNC turning machines, are machine tools that rotate a bar of material, allowing the cutting tool to remove material from the bar until the desired product is remaining. The material itself is secured to, and rotated by, the main spindle, while the cutting tool can be moved along multiple axes. Like the CNC milling machine, this is operated using CAD software.

▲ A CNC centre lathe

CNC plasma cutters

CNC plasma cutting refers to the cutting of metals using a plasma torch controlled from a computer. Plasma cutters operate by forcing a gas or compressed air at high speeds through a nozzle. Once an electric arc is introduced to the gas, ionised gas or plasma is created.

CNC water-jet cutters

A CNC water-jet cutter uses a high-pressure jet of water to cut various materials. Water alone can be used to cut soft materials like wood and rubber. For harder materials, the water can be mixed with an abrasive substance (like garnet or aluminium oxide).

▲ A CNC plasma cutter cuts using a plasma torch.

▲ A CNC water-jet cutter

Task 3.3

1. What does CAD stand for?
2. List two advantages of CAD.
3. List two disadvantages of CAD.
4. What does CAM stand for?
5. List two advantages of CAM.
6. List two disadvantages of CAM.
7. What does CNC stand for?
8. Identify the CAM used in the pictures below.

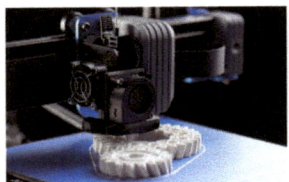  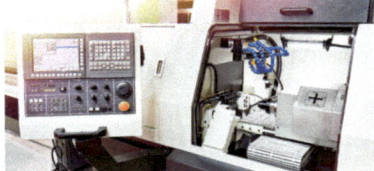 

Health and safety

Introduction

This part of the chapter covers health and safety in the workshop environment. All engineers have to understand and use a large range of tools and equipment during their career, whether it is starting out and going through training, or working in an office environment and specifying different pieces of equipment to be used for different processes. Engineers also need to understand the health and safety procedures that have to be put in place in working environments to ensure safe working practices for the people involved.

When working on a machine that spins sharp/heavy pieces of metal at 2000 RPM, welding steel at 10 000 °C, or even producing toxic fumes from acid-etching circuit boards, you need to make sure you can walk away from the process and equipment perfectly safely, with a quality outcome product. This is why engineers need to understand and apply health and safety procedures for each task that they perform.

▲ Various items of personal protective equipment

Personal protective equipment

Personal protective equipment (PPE) is what we call items of clothing or guards that protect us from harm when using dangerous machinery.

Safety goggles/glasses

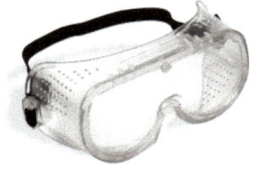

▲ Safety goggles or glasses are a protective barrier and ensure the safety of your eyes.

Safety goggles or glasses should be used when operating large machinery or when conducting tasks where there is a potential for wood shavings or swarf to flick into your eyes.

Aprons

Aprons are worn in workshops to protect clothing from being ruined by chemicals such as cutting compound or engineer's blue. An apron also acts as an extra protective layer when using any corrosive or harmful chemicals or finishing products.

▲ An apron is worn to protect clothing in a workshop environment.

Hair ties
Hair ties are essential in the workshop if you have long hair. Hair ties prevent your hair from becoming an obstacle when using any large or dangerous machinery. It would be best to wear a hair tie as soon as you enter the workshop to ensure you are working safely.

▲ Hair ties

Welding helmet
A welding helmet is also commonly referred to as a hood. This piece of equipment is vital when conducting forms of welding, such as brazing or arc welding, as it protects your eyes and skin from sparks. However, the main purpose is to protect your eyes from potentially vision-damaging ultraviolet and infrared rays.

Gauntlets
Gauntlets are also known as welding gloves and are used during soldering or extreme heat work, such as welding. They are made from non-conductive material which allows the heat to dissipate, therefore protecting the engineer from burning, sparks and potential abrasions.

▲ A welding helmet (hood) and a pair of gauntlets or welding gloves

Ear defenders
Ear defenders can come in many forms, such as earplugs, earmuffs and canal cups. The main aim of ear defenders is to reduce the high level of sound reaching your ears from machines in the workshop and so protect your ears from damage.

Safety boots
Safety boots are fitted with a protective steel cap that means if a large or heavy item is dropped on your foot, it will not injure your toes. Wearing safety boots also offers protection from punctures, cuts and burns. The soles of safety boots provide grip when walking on slippery or uneven surfaces, helping to prevent falls which can lead to more serious problems.

▲ A workshop can be a very noisy place so ear defenders are essential.

Respiratory protection
Wearing respiratory protection is essential when engineers are completing tasks such as sanding, which produces large amounts of dust particles. **Respiratory protective equipment (RPE)** is a particular type of PPE, used to protect the individual wearer against the inhalation of hazardous substances in the workplace air.

▲ Safety boots

Risk assessments

A **risk assessment** is an analysis of the risks involved when using equipment or performing a process.

Imagine making a cup of tea. Do you need to be aware of any dangers? What about the mixture of water and electricity? Would you be aware of the heat generated by the boiling water? What would you do to keep yourself safe and ensure you end up with a nice cup of tea? Identifying risks and putting procedures in place to keep yourself (and others) safe is part of a risk assessment.

The five-step risk assessment
One of the most common (and industry-recognised) methods for preparing risk assessments is the five-step method:

Step 1. Identify the hazards.

Step 2. Identify who may be harmed and why.

Step 3. Evaluate risk and choose precautionary control measures.

Step 4. Record (write down) your findings.

Step 5. Review and update when needed.

▲ RPE prevents the wearer from inhaling hazardous substances in the workshop.

Top tip

Risk assessments can be produced for individual machinery, equipment or for projects.

Following is an example of a five-step risk assessment for using a bench/pillar drill.

Risk assessment for bench/pillar drill (school workshop)		
Identify the hazards	**Who may be harmed and why?**	**Evaluate risk and choose precautionary control measures**
1 Workpiece spinning on drill bit.	1 User cutting hand on rotating workpiece.	1 Use machine guard to reduce risk of rotating workpieces contacting user.
2 Workpiece flying off machine.	2 User being hit by flying workpiece.	2 Use machine guard to reduce risk of user being hit.
3 Swarf flying up.	3 Swarf flying into user's eyes.	3 Wear goggles to reduce risk of eye injury.
4 Long hair/loose clothing tangling in rotating parts.	4 User being dragged into rotating parts.	4 Wear apron and tie down loose hair/clothing to reduce risk of entanglement.
Note:		
All manufacturer guidelines should be used when this machine is in operation.		

Having completed the first three steps, the fourth would be to produce a document (e.g. the above table) that would be kept on record and readily available for the users of the machine (bench/pillar drill) to access and read. Step 5 would be done if there are any changes made to the machine (or law). Quite often, risk assessments are reviewed and updated on an annual (yearly) basis to make sure the procedures used are still up to date.

Identify the hazards

An important aspect of a risk assessment is determining what could be potential hazards in the workshop or workplace. It can be a good idea to start by walking around the workshop and looking for things with the potential to cause harm. What processes, activities or chemicals could cause harm or injury to a person's health?
- What jobs are being done?
- Is the equipment and machinery being used properly?
- Are there any harmful chemicals or substances in use?
- Can you see any unsafe work practices?
- Is the workshop in a good general state?

A quick look at the accident book might help you identify less obvious hazards or things that have been responsible for near misses. It can also be a good idea to consider less obvious operations, such as maintenance and cleaning.

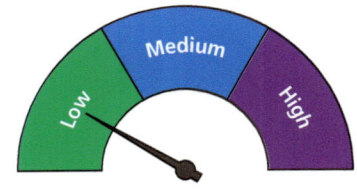

▲ *Identifying risk*

Key Term

Risk matrix A diagram used to define the level of risk by considering the category of probability or likelihood of an incident occurring against the category of consequence severity.

Identify who is at risk

For each hazard identified, the risk assessment must be clear about who might be harmed. Consideration should also be given to your peers and teachers and, in fact, anybody else who may be in the workshop. Think about people interacting with the hazard either directly or indirectly. For example, a student varnishing their completed project is directly exposed to solvents at close proximity, while other students in the vicinity are inadvertently and indirectly exposed.

Evaluate the risks and consider whether they can be eliminated

After identifying the hazards and those at risk, the next step is deciding how likely it is that harm will occur. In other words, determine the level of risk and evaluate any existing control measures.

The easiest way to do this is by using a **risk matrix** to define the level of risk (see tables below). This is a diagram used to define the level of risk by considering the category of probability of an incident occurring against the category of the severity of the consequence of the incident. Each category is assigned a score, usually between 1 and 5. By multiplying the scores for probability and consequences together, a risk score will be generated for each activity.

Consequences

Score	1	2	3	4	5
Description	Insignificant	Minor	Moderate	Major	Catastrophic
Example	Minor injury, no first aid required	Harmful injury (first aid required, under 3 days recovery time)	Serious injury, medical assistance required (injury must be reported)	Major injury, urgent medical assistance required	Fatality

Likelihood

Score	1	2	3	4	5
Description	Rare	Unlikely	Possible	Likely	Almost certain

Risk matrix

Consequences/impact						
Catastrophic	5	5	10	15	20	25
Major	4	4	8	12	16	20
Moderate	3	3	6	9	12	15
Minor	2	2	4	6	8	10
Insignificant	1	1	2	3	4	5
		1	2	3	4	5
		Rare	Unlikely	Possible	Likely	Almost certain
		Likelihood/Probability				

Consider what is already being done and any risk controls in place. The first option for minimising risk would be to try to remove the hazard. For example, if the milling machine guard is broken and swarf can get through, would you continue to use the milling machine or would you remove the risk by using a hacksaw and file to create the desired angle of your workpiece?

If the hazard cannot be eliminated, consider how the risks can be controlled so that harm becomes unlikely.

Signage

Signage is the word used for all the signs that you may see in a workshop environment. Knowing how to translate and understand the signs in a workshop is vital when dealing with potentially dangerous equipment and processes. By ignoring signage, you could be in danger of harming yourself or others. Do you know the signs used to warn you that someone may be arc welding in your workshop? What would happen to your eyesight if you did not understand the signs and just popped your head into a welding booth to say hello? Could you lose the use of your sight temporarily or even permanently? This is why all engineers should be aware of the signage that is displayed in any workshop environment. Safety signs tend to come in different shapes and colours that have specific meanings attached to them. The following table summarises what is meant by the different shapes and colours used.

Sign	Meaning	Shape	Colour
	**Mandatory sign:** specific instruction on behaviour	Round	White border, blue background, white pictogram
	Warning sign: giving warning of hazards or danger	Triangular	Black border, yellow/orange background, black pictogram
	Prohibition sign: prohibiting behaviour and/or actions	Round	Red border, white background, black pictogram
	No danger: information on emergency exits, first aid, emergency stop, etc.	Square or rectangular	White border, green background, white pictogram

> **Top tip**
>
> You can find a list of common signs that you may find in a workshop environment on the following two websites:
> - Freesignage UK: www.freesignage.co.uk
> - Online Sign: www.online-sign.com

Quite often, signs will come with some instruction in the form of text to reinforce the message.

Some areas of work, such as construction sites or workshops, may have multiple instructions that have to be applied at the same time. The following figure is an example of a sign with multiple instructions for visitors arriving at a construction site.

Mandatory signs

The following section gives examples of some mandatory signage that may be found in working environments.

PPE

3 Using engineering tools and equipment

Other mandatory signs
Warning signs

Perygl / Caution | Rhybudd sioc drydanol / Caution electric shock risk | Perygl paladr laser / Caution laser beam | Rhybudd gwasgu dwylo / Warning crushing of hands | Perygl nwy cywasgedig / Danger compressed gas | Lefelau sŵn uchel / High noise levels

Prohibition signs

Peidiwch â chyffwrdd / Do not touch | Peidiwch â chyffwrdd pan fydd yn symud / Do not touch when in use | Peidiwch ag iro na glanhau pan fydd yn symud / Do not oil or clean when in use | Peidiwch â rhedeg / Do not run | Dim fflamau noeth / No naked flames | Dim bwyd na diod / No food or drink

No danger signs

 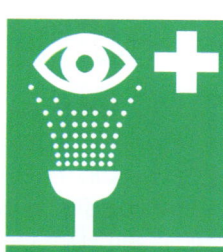  

Botwm argyfwng / Emergency stop | Golchi llygaid mewn argyfwng / Emergency eye wash | Safle cymorth cyntaf / First aid station | Allanfa argyfwng / Emergency exit

Using data sheets

When using tools and machinery you will be given instruction on their correct and safe use by your tutor. However, when using machinery such as pillar drills, milling machines and centre lathes, you will have a number of variables, such as cutting tool sizes or diameters and types of materials, that would change depending on what job, project or operation is being performed. Because all these aspects tend to change frequently, the machine settings would also need to change to match the current job.

For example, one day you could be milling a small aluminium bracket on the vertical miller with a 5 mm diameter cutting tool, then the following day you could be trying to mill a large slab of stainless steel with a 12 mm diameter cutting tool. In this situation you would have to adjust the speed of the cutter (RPM) as well as how fast (feed) the cutting tool would mill the work.

This is where good engineers use data sheets/charts that are either developed by other machine users or come direct from the manufacturer. There are mathematical formulae that can be used to determine the speed and feed rates of all machines, depending on material properties and cutting tool sizes; however, there are many data guideline sheets that will give you a good idea of the speed and feed rates needed for your current job.

Following is an example of a data sheet chart for a pillar drill. See how it has 'Extra notes' at the bottom, giving advice and guidelines on specialist operations such as counterboring and countersinking.

Guideline RPM for pillar drill					
Drill bit diameter		**Material**			
Inches	Millimetres	Aluminium	Brass	Mild steel	Stainless steel
$\frac{1}{4}$	6.35	1200	1200	750	500
$\frac{3}{8}$	9.5	900	600	500	375
$\frac{1}{2}$	12.7	600	500	250	250
$\frac{5}{8}$	15.8	360	400	200	150
$\frac{3}{4}$	19	300	300	150	120
1	25.4	180	200	100	80

Notes:

Counterbore bits and operations should run at 200 RPM maximum.

Countersink bits and operations should run at 300 RPM maximum.

By utilising these data sheets every time you set up a machine for a job, you will reduce the risk of the machine breaking, the cutting tool breaking or the workpiece being ruined. Above all, you will ensure the safety of the operator ... you.

COSHH

Key Term

COSHH Control of Substances Hazardous to Health. A set of regulations drawn up by the Health and Safety Executive (HSE) in 2002 to provide guidelines on the safe handling of hazardous substances.

All engineers need to be aware of the **COSHH** regulations. In a workshop environment, you will be working with a number of substances that could be hazardous to your health and it is essential that you understand how to handle and store these substances safely. Substances that fall under the remit of COSHH include:

- chemicals
- products containing chemicals
- fumes
- dusts
- vapours
- mists
- nanotechnology
- gases and asphyxiating gases
- biological agents (germs)
- germs that cause diseases such as leptospirosis or legionnaires' disease and germs used in laboratories.

(Source: HSE (2019) *What is a 'substance hazardous to health'?*, www.hse.gov.uk/coshh/basics/substance.htm)

If packaging has any of the hazard symbols on it then it is classed as a hazardous substance.

▲ *Dangerous fumes come under the remit of COSHH.*

If any of the substances listed above are in a workshop you are working in or you are going to use any of the listed substances, then you must follow the government guidelines on their correct use and storage.

As you can see, not all the COSHH substances listed would be found in a workshop environment. However, here are some examples of substances that could be found in a workshop that would have to adhere to COSHH guidelines:

- paint
- varnish
- undercoat
- thinners
- solvents
- adhesives
- wood finishes (wax, etc.)
- acids
- fumes (welding, spray booths, etc.)
- dust particles (linisher/belt sanders/sanding).

▲ *The use of paints in a workshop would need to adhere to COSHH guidelines.*

Other areas of safety guidelines in a workshop environment

There are a number of organisations that deal with health and safety in workshop environments. These organisations employ experts in their field and develop guidelines (not rules) on safe working practices from the setting up of workshop spaces (distances between each machine, etc.) to how to safely use and maintain individual pieces of equipment and machinery such as milling machines and table saws. Two of the most widely used organisations in schools, colleges and university workshops are:

- DATA (the Design and Technology Association)
- CLEAPSS (Consortium of Local Education Authorities for the Provision of Science Services).

 Top tip

To find the guidelines for each of these substances you can visit the government website www.hse.gov.uk/coshh/. However, your tutor should have procedures in place to ensure safe working practices are available to you.

 Top tip

Any further information or guidelines on health and safety in a workshop environment can be found on the websites of either of these two organisations:
- www.data.org.uk
- www.cleapss.org.uk.

Task 3.4

What do the colour and shapes of the signs mean?

Task 3.5

Choose a machine that might be found in a school workshop, such as a pillar drill or centre lathe, and complete a risk assessment for that machine, using the headings below.

Identify the hazards	Who is at risk?	What is the harm?	Activity taking place	Control measures required	Additional information
Employees and learners should be made aware of the following hazards.					

Implementing engineering processes

In this chapter you are going to:
- learn how to correctly identify processes needed to complete a project
- understand how to join (fabricate) materials effectively
- understand how plastics can be formed with different moulding processes.

To successfully achieve the objectives, you should have access to a workshop environment with enough available fabricating (making) processes to demonstrate your engineering skills. (Your school or college should provide these facilities.)

This chapter will cover the following areas of the WJEC specification:

Unit 1 Manufacturing engineering products: Unit 1.4 Implementing engineering processes			
• 1.4.1 Apply a range of engineering processes	• 1.4.2 Work with a range of materials	• 1.4.3 Evaluate the quality of engineered products	• 1.4.4 Evaluate own practices and processes
Unit 3 Solving engineering problems: Unit 3.3 Understanding methods of preparation, forming, joining and finishing of engineering materials			
• 3.3.1 Describing engineering processes	• 3.3.2 Describing applications of engineering processes	• 3.3.3 Safe working practices	

Introduction

In the previous chapter, we discussed how different tools and equipment are used, the technical information needed to identify them and what processes they can be used for. In this chapter, the focus will be more on the industrial processes that engineers should know about so they can specify what process might be needed when undertaking a project. Engineering processes are ways of fabricating (putting together and shaping) different materials and this chapter will show you how different processes can be used to form different materials into products or parts of products.

Marking-out

If you were drawing on a piece of paper, you would use pencils, erasers, compasses and set squares. It is much more difficult to write and mark on a metal such as steel than it is to write or draw on paper. In this instance, as we have seen in previous chapters, the metal to be worked on would be known as the **workpiece** and the process of drawing onto the workpiece is known as **marking-out**.

There are two main ways of marking-out: flat and vertical. In **flat marking-out**, the workpiece is laid flat on a surface; in **vertical marking-out**, the workpiece is held in a vertical position. The following table details the correct marking-out tools for different materials

Process	Wood	Metal	Plastics
Lines	Pencil	Scriber	Felt-tip pen
Lines at right angles to an edge	Carpenter's try square	Engineer's try square	Engineer's try square
Lines parallel to an edge	Marking gauge	Odd-leg callipers	Odd-leg callipers
Marking for a mortise	Mortise gauge	N/A	N/A
Marking a circle	Pair of compasses	Dividers	Dividers
Marking the centre of a hole	Pencil	Centre punch	Felt-tip pen
Marking an irregular shape	Template	Template	Template

Flat marking: metal

If surfaces that are supposed to be flat are damaged, or there is dirt between the surfaces, then the marking-out process will not be accurate. It is, therefore, important to protect and clean all the marking-out equipment before it is used. The workpiece must also be cleaned with a cloth to remove oil and grease.

The following list details the procedures involved in the flat marking-out of a metal surface.
- Use **marking blue**, a dark blue runny ink, to paint the surface of the metal. This will dry very quickly. Once the marking blue is dry, the workpiece is ready to be marked.
- A **scriber** is held like a pen and has a sharp steel point. It is used like a pen to mark lines onto steel.
- Lines drawn upon engineering products will be either straight or curved. A **rule** is used for straight lines. The rule is placed on the workpiece and the scriber is used to mark the line.
- When a line needs to be at a right angle to the edge of the workpiece, an **engineer's try square** is used. The engineer's try square is held in place by hand and the line is scribed with the scriber.
- If an angle other than a right angle is required, a **combination square** may be used. This is used in a similar way to an engineer's try square but it can be set at any angle.
- **Odd-leg callipers** are used to produce a straight line parallel to the edge of the workpiece. One edge of the callipers has a step and the other has a sharp point. The sharp point does the work of a scriber and marks a fine line on the workpiece.
- **Dividers** are used on metal in the same way as compasses are used on paper. There are two points, one on each end of the dividers. One point is placed in a small indented hole on the workpiece and the dividers are then used to produce a circle or arc on the surface of the metal.
- The small indentation needed in order to use a pair of dividers on metal is produced by a **centre punch**. A centre punch is similar to a scriber but much thicker and heavier. It is hit with a hammer to create a small indentation in the workpiece. This indent may be used to locate the dividers in order to produce the circle or arc. A centre punch could also be used to produce a series of small indents along the scribed line. If the blue marking is worn from the surface of the workpiece, the small indents produced by the centre punch can still be used as a guide. The indentation produced by a centre punch could also be used to locate a drill. This would require a bigger indentation than for the purposes described above. The drill bit is located in the indent and the hole is drilled in that position.

▲ *Flat marking on wood with a steel rule*

▲ *Flat marking on metal*

▲ A cast iron surface plate

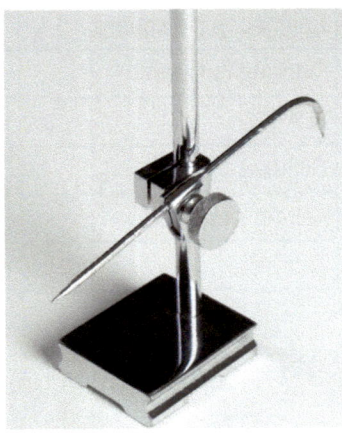

▲ A scribing block

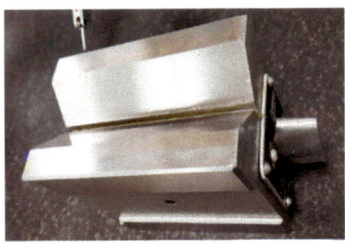

▲ V-blocks and clamps

> **Key Term**
>
> **Fabricate** To manufacture something from different parts; to shape and join materials to create a product.

Vertical marking: metal

Often it is better to stand the workpiece up vertically in order to mark it out. This will require some special marking-out equipment.

- A **surface table** is a perfectly flat table made of cast iron, steel or graphite. Sometimes a smaller flat plate is used known as a **surface plate**.
- The surface table or plate must be protected from damage and is usually covered when not in use. Vertical marking-out must be done on either a surface table or surface plate, as these are guaranteed to be flat, rather than a general workbench or desk. **Parallels** are perfectly flat lengths of bar which sit upon the surface table. The workpiece sits on top of these, which makes the process of marking-out much easier.
- An **angle plate** is an L-shaped device with an angle of exactly 90 degrees. The workpiece rests against the angle plate, which ensures that it is perfectly vertical. The workpiece can be either held against the angle plate or clamped in place using slots in the angle plate.
- A **scribing block** holds a scriber in place. It has a flat base so it can sit perfectly flat on the surface table. When the **surface gauge** moves left to right, the point of the scriber always remains at exactly the same height.
- If a round or cylindrical bar of metal is to be scribed, it can be held in place by a **V-block**. The bar sits in the V-block and may be held firm by a small clamp.

Joining materials

Engineers need to understand how to join materials. New products are mainly made up or **fabricated** from different parts that have been joined in some way. Knowing how to join the different parts is a much-needed skill.

You have probably joined or fabricated different materials or components already. There are many different methods of joining, including adhesives, tapes, screws, nuts and bolts, joints, hinges, welding and soldering.

The joining of parts/materials is generally broken down into two parts:
- **permanent** joints/fixings
- **temporary** joints/fixings.

Permanent joints/fixings

Permanent joints are exactly that ... permanent. They should not come apart during the product's life. They include:
- soldering
- brazing
- welding
- rivets
- adhesives.

The following section covers engineering processes used to permanently join materials together. This is also known as 'fabricating' products.

Soldering

Soldering is a process that is used to join metal pieces together. It is commonly used in the plumbing industry and to join the components in circuits boards.

Solder is an **alloy** of two different metals that, when combined, create a soft metal with a very low melting temperature. In the past, solder was made from tin and lead; however, due to restrictions on the use of lead in consumer products, a lot of solder is now made from tin, copper, zinc or silver. (Industrial-use solder uses silver.)

Due to its low melting temperature, solder can be heated and melted to form around other metals without the other metals melting. This acts as a kind of metal 'glue'.

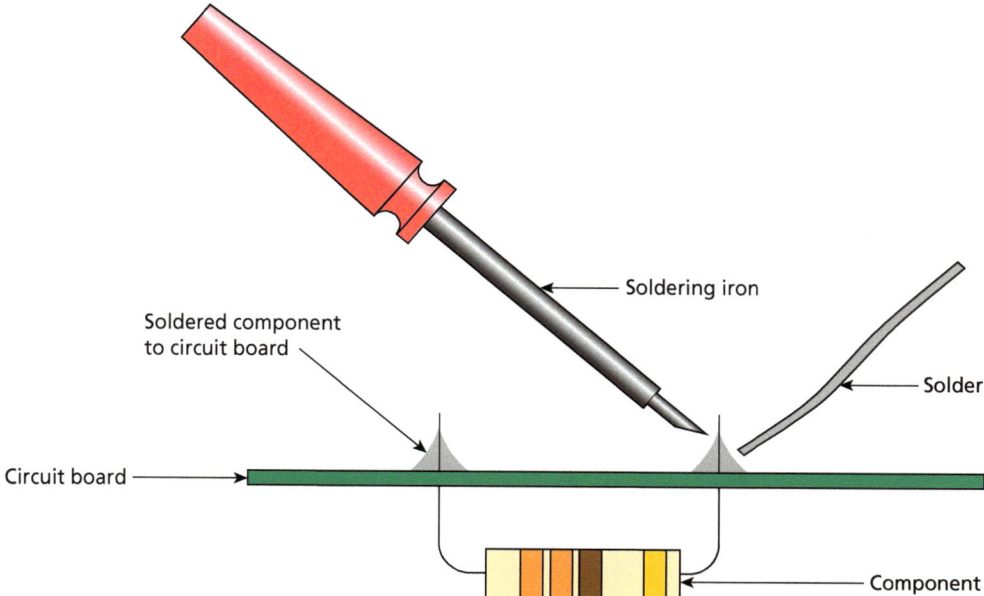

▲ An example of the soldering process and how a component (in this case a **resistor**) can be permanently joined to a circuit board.

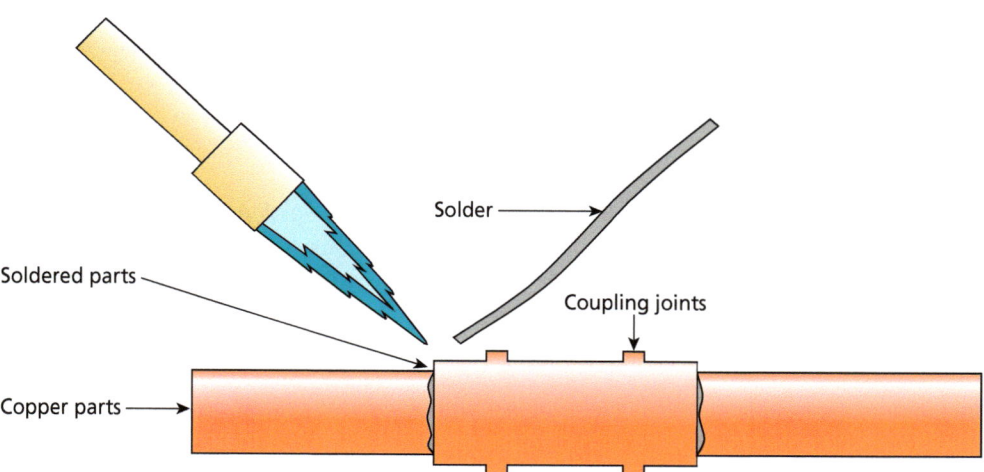

▲ An example of how plumbers could solder copper pipes together using heat from a blowtorch.

Brazing

Brazing is another process that can be used to join different metals together. It is good for joining steel (mild/stainless) to other metals as well as steel to steel. In brazing, brass is used as the **sacrificial metal**; the brass is melted and joins the other metals as it cools. Brass is a lot stronger than solder and therefore creates a much stronger joint. Brass also has a higher melting temperature than solder (but still lower than that of steel) and needs a higher temperature flame. Quite often an **oxy-acetylene torch** is used because of the high temperatures it produces (approximately 3500°C).

As with soldering, flux is used to clean the area to be joined. The melted brass then flows where the flux has been applied using **capillary action**.

Top tip

Flux is a chemical that can be used to clean the surface of joints before soldering. The flux allows the solder to have good contact with the metals to be joined and enables the solder to flow into the areas that have had flux applied.

Key Term

Resistor An electrical component that can be used in a circuit to reduce/slow down the current in it.

Key Terms

Capillary action When liquid flows through very narrow spaces. In the brazing example, the **molten** metal is the liquid flowing through the space between two touching pieces of steel.

Molten When a solid is heated to the point at which it becomes a liquid, it is said to be molten.

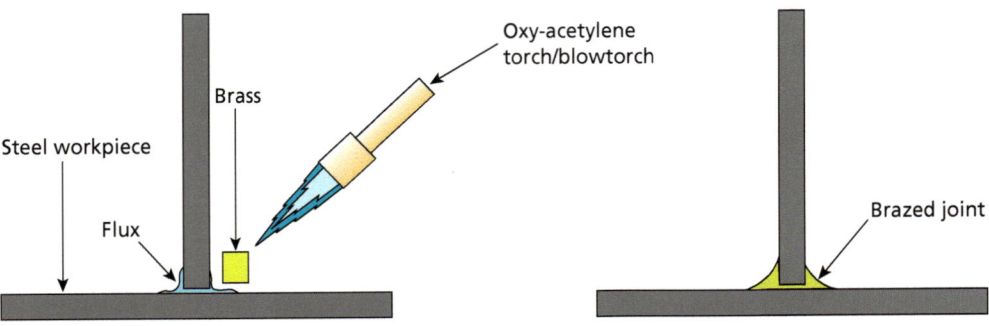

▲ *An example of the brazing process where two pieces of steel are permanently joined.*

Instead of placing small amounts of brass in the area to be joined, you can also use 'filler rods' that are consumable metal rods made from brass.

▲ *A filler rod*

▲ *A brazing hearth*

▲ *A welding bay*

▲ *A welding table*

Brazing hearth

A brazing hearth is similar to a workbench, as it is a solid and sturdy piece of equipment, but it has the addition of robust sides that act as a heat shield. The brazing hearth contains a compressor which pressurises air and gas so that it is forced out of the nozzle of a gas–air torch.

Welding

Welding is another process that joins metals together to create a strong permanent bond. The difference between welding and soldering or brazing is that in welding it is the metals to be joined that are heated, rather than a sacrificial metal such as solder or brass.

Welding bay

A welding bay or area will consist of a welding curtain and a welding table. The welding curtain will cover the area where welding is taking place. This curtain is made of a heat-resistant material that is designed to be a barrier to protect against sparks and the welding light that could cause damage to eyes.

Welding table

Welding tables are designed specifically for the process of welding, with holes positioned to hold welding projects in place with clamps, jigs and stops. The material is held tightly in place while it is being welded and assembled.

Different types of welding

MIG (metal inert gas) welding

MIG welding is used to permanently join steel to steel. It uses an electric current to create a powerful **electric arc** between the steel joint you are creating and a consumable steel wire (also known as an **electrode**). The intense heat of the electric arc melts the workpiece and the consumable steel wire into a molten pool where they join and create a weld. MIG welding also uses a gas to act as a flux and clean the joint as you weld. Both the gas and the consumable wire are fed through the MIG torch. MIG welding is generally used on smaller projects with thinner materials.

> **Key Terms**
>
> **Electrode** An electrical conductor that is generally used to make contact with a non-metallic part of a circuit. In MIG and arc welding, the electrode is the sacrificial metal wire or rod.

4 Implementing engineering processes

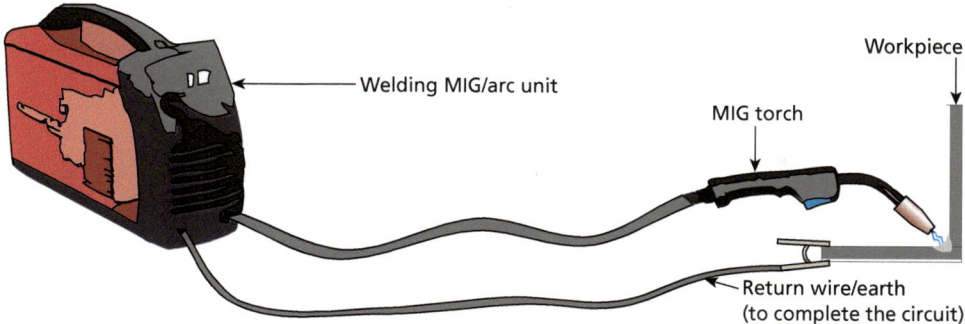

▲ An example of a MIG welding unit (left) and two pieces of steel being MIG welded (right).

Stick welding

Stick welding (also known as metal arc welding) works on the same process as MIG welding; they both use electricity to create a powerful electric arc between an electrode and a workpiece that is hot enough to melt steel. Instead of using a wire, stick welding uses a consumable rod, held with a rod holder. The centre part of the rod is mild steel and it is covered with flux. As you move the rod across the workpiece, the rod is consumed and gets smaller. Stick welding is generally used for medium- to larger-sized projects with thicker materials.

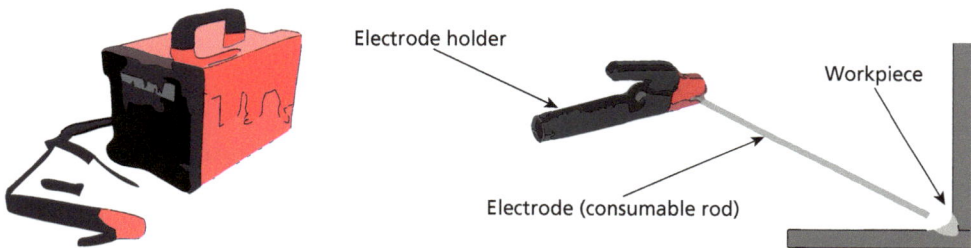

▲ An example of an arc welding unit (left) and two pieces of steel being arc welded (right).

Oxy-acetylene gas welding

Oxy-acetylene gas welding is most commonly used to fabricate (shape/join) steel. It uses a mixture of oxygen and acetylene gas to produce an extremely hot flame (approximately 3500°C) to melt metals. The flame is so hot, the steel turns into molten pools of metal. A filler wire can also be used to 'fill' as it melts with the heat. The hot steel from the filler wire and workpiece mix together and then join to form a single piece.

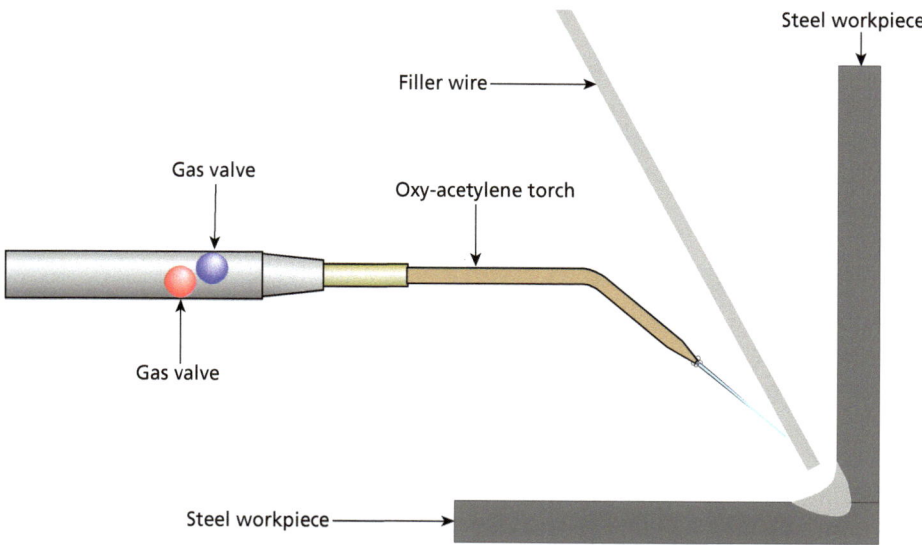

▲ Two pieces of steel being permanently joined using the oxy-acetylene gas welding process.

79

WJEC Level 1/2 Vocational Award Engineering (Technical Award)

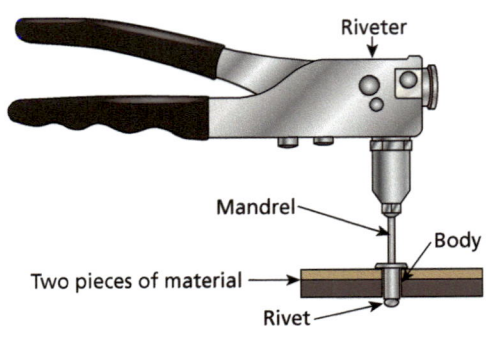

▲ Riveting can be used to permanently join two thin pieces of metal.

Rivets

Pop riveting is an engineering process that is used to join two thin pieces of metal together. The rivet is made up of a mandrel, a rivet body and a rivet head.

Before starting, it is important to measure the diameter of the rivet body. A hole of the correct size is then drilled in the workpieces that need to be joined and a rivet inserted. A rivet gun (riveter) is then placed over the mandrel and the handled is squeezed as many times as necessary to extract the mandrel from the rivet body. The two workpieces should now be firmly clamped between the rivet head and the deformed end of the rivet body. At the end of the process, the mandrel can be discarded.

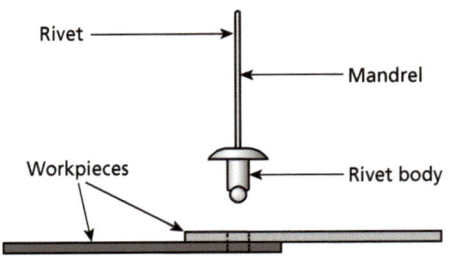
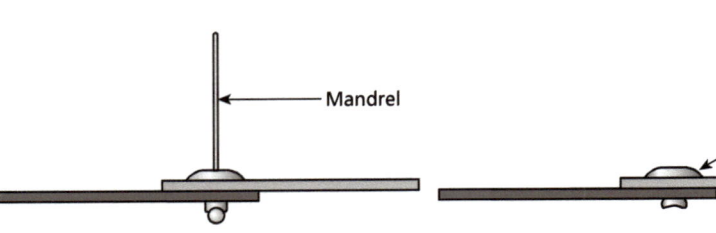

▲ The three-step process involved in pop riveting.

> **Key Term**
>
> **Mandrel** The long, cylindrical part of a rivet.

Adhesives

There are many types of adhesives available to engineers that can be used for multiple different projects and purposes. When an engineer glues two pieces together, using adhesives, the outcome will be a permanent joint. Adhesives are common in schools and for DIY projects.

Superglue

Superglue is an extremely strong adhesive that is used to join materials together permanently. Superglue is available in a number of different strengths.

Wood glue

Wood glue is an adhesive used to bond pieces of wood together tightly. Once glued, a permanent joint is created, making it extremely difficult to dislodge these joints.

Epoxy resin

Epoxy resin is one of the strongest adhesives available, and therefore one of the more common adhesives used in engineering. It comes in two tubes, the contents of which need to be mixed together: one is the resin and the other is a hardener. When these are mixed together, they harden quickly and make an extremely strong, permanent bond between materials.

▲ Superglue

▲ Wood glue

▲ Epoxy resin comes in two tubes that must be mixed together.

Temporary joints/fixings

Temporary joints can last a long time but they are are not designed to be permanent and do come apart eventually. They include:
- temporary adhesives
- tape
- screws
- nuts and bolts
- knock-down fittings (components used to make flat-pack furniture).

4 Implementing engineering processes

Temporary adhesives

Hot glue gun
A hot glue gun is a hand-held device that uses a heating element to heat and melt a solid glue stick. Once the adhesive has melted, it can be directed out of the gun's nozzle and onto a given object. Like PVA, it does not provide a strong bond between materials.

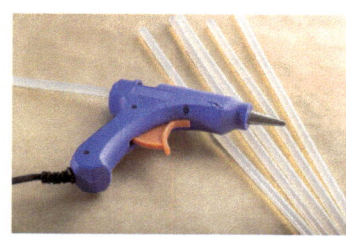

▲ A hot glue gun

Tape
Tapes come in a reel and can be any one of a number of varying styles, including electrical tape, sticky tape, masking tape and duck tape. Tape is only used to fasten or hold objects short term.

▲ Various styles of tape

Screws
Screws are available in varying sizes and lengths. The most common uses of screws are to hold objects, such as pieces of wood, together and to position objects. There is often a head on one end of the screw that allows it to be turned. The head is usually larger than the body of the screw. Screws can be applied and removed from materials using a screwdriver or an electric drill.

Nuts and bolts
Nuts are used in conjunction with a bolt to fasten multiple parts together. The two partners are kept together by a combination of their threads. They can be screwed into and out of materials easily.

▲ Knock-down fittings are often used for flat-pack furniture.

Knock-down fittings
Knock-down fittings are very common fixtures in flat-pack furniture. This is because they can be put together easily, normally requiring only a screwdriver, a drill, a mallet/hammer and other basic tools. Like previous fixings, they can be unscrewed easily from the product if necessary.

Key Term

Planish To flatten, smooth or polish (metal) by rolling or hammering it.

Shaping materials

The following processes involve shaping materials into usable forms. When shaping metals, it is worth noting that after each 'forming' process, the end product still has to be finished.

Planishing

In the process of **planishing**, many light blows are used to smooth metal which has already been formed by some other means.

A **doming block** and **punches** can also be used to create different curves and shapes with the planishing process.

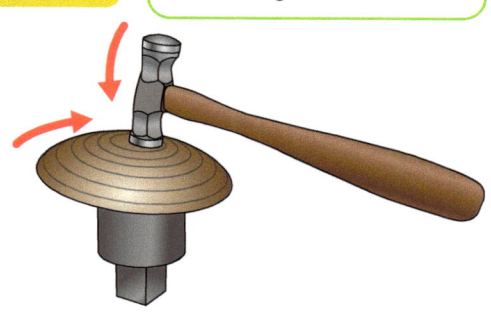

▲ Planishing

Task 4.1
In your notebook, identify the different joining processes and describe where the process might be used.

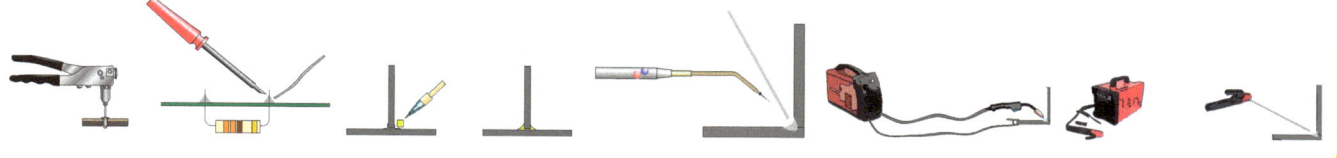

WJEC Level 1/2 Vocational Award Engineering (Technical Award)

Key Terms

Malleable Pliable, easy to shape without cracking or breaking.

Billet Also known as a billet of metal, a billet is a piece of metal of a certain size that is shaped by the forging process.

Sprue A hollow channel through which molten material (metal or plastic) is poured into a mould. It is the primary feed channel for the molten material to enter the mould cavity. The sprue allows the molten material to flow from the source to the mould, filling the mould cavity and creating the desired product. In the context of the final product, the sprue is considered waste material because it is typically removed from the finished product. However, sprues can often be reused.

Runners A channel that guides the molten material from the sprue to the individual parts or cavities within the mould. It branches out from the sprue to distribute the material. The runner system helps evenly distribute the molten material to different parts of the mould. It ensures that each part receives the required amount of material for proper formation. Like the sprue, the runner is also considered waste material in relation to the final product but they can be reused.

Forging

Forging is a process of joining and/or shaping metals. In forging, force is used to bond workpieces together or change the form of a workpiece into a desired shape. The most common image people have of forging is a blacksmith using a hammer and anvil to shape heated metal. When metal is heated, it becomes more **malleable** and is easier to shape. If the metal is hot enough, it can also be bonded with the use of force. (Molten metal pools together to create a weld, as in the welding process.)

▲ *The process of forging*

To heat the metals you intend to work on, you would need a forge. Forges can be fuelled by either coal (a more traditional type) or gas (more modern). Gas forges allow the user to control the temperature more accurately and therefore allow the metals to be forged at the correct temperatures more consistently.

Drop forging is an industrial process where the force needed to forge two workpieces together is provided by a machine. The drop forge also has a die (upper and lower) in the shape of the product you are going to forge. The die is essentially a mould. A heated piece of metal, called a **billet**, is placed in the die. The upper die is dropped with force onto the billet and lower die to create the shape needed.

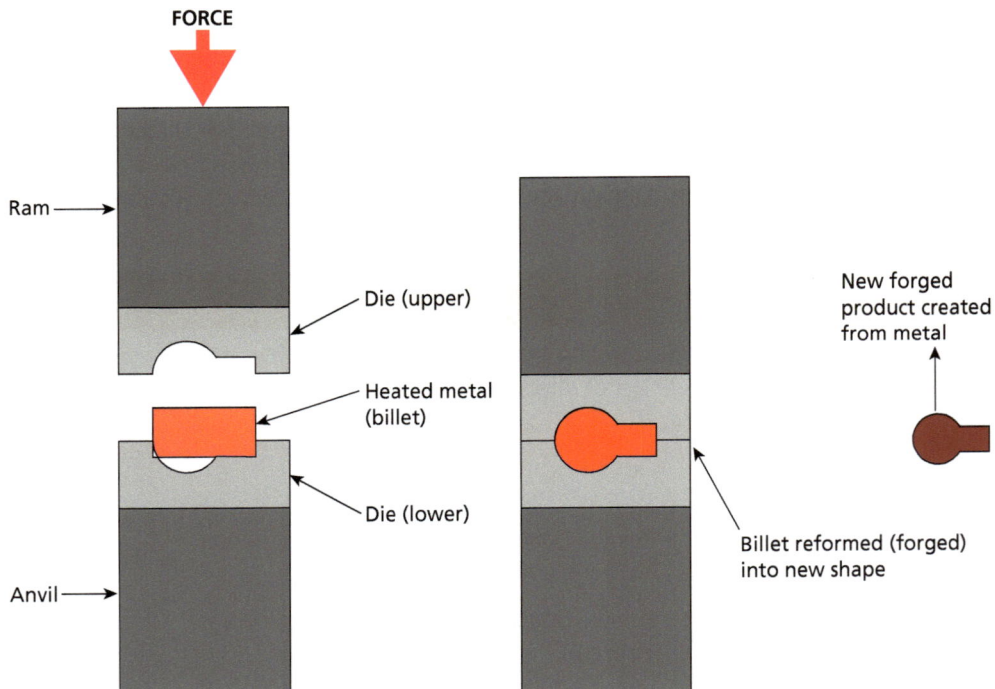

▲ *The drop forge process*

Casting

Casting is a process in which metals are heated until molten and then poured into a mould. The metal is left to cool and the result is a metallic object created in a desired shape. Most have to be finished before the final object is complete. Finishing could include cutting off the **sprue** and **runners**, filing, milling, grinding, drilling and/or polishing.

Sprues and runners are channels used in casting processes to facilitate the flow of molten material into the mould. They are considered waste material in relation to the final product, but they can be recycled and reused in an effort to minimize waste.

4 Implementing engineering processes

Sand casting

Sand casting is an old process that is still commonly used today. The process involves taking an existing, permanent shape (**pattern**) and then packing sand around that shape to create a temporary, **expendable mould**. The pattern is then removed to leave a cavity into which molten metal is poured. Once cooled, the sand mould can be broken apart to reveal the solid metal object in the shape of the pattern. The sand can then be reused for another mould.

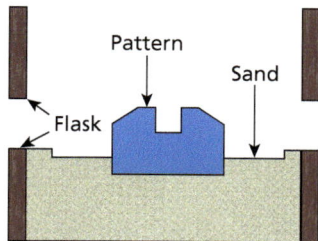

1 Sand and a permanent pattern are placed into the bottom half of a flask (sand mould holder).

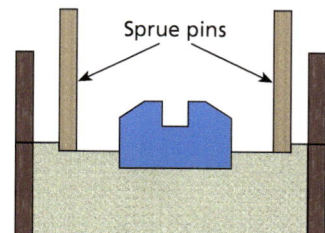

2 Sprue pins are placed to create the riser and the top half of the flask is attached.

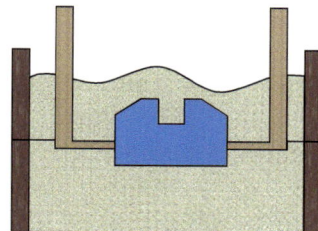

3 The rest of the flask is filled to create the mould.

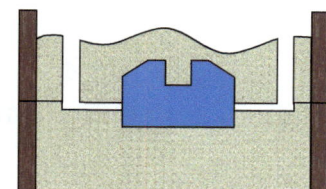

4 Sprue pins are removed.

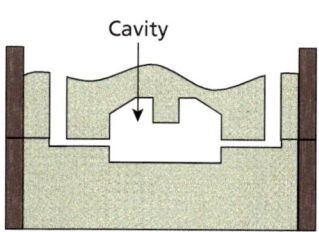

5 The pattern is removed to create the desired cavity shape.

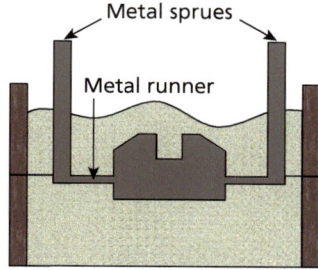

6 Molten metal is poured in to create the desired product.

7 The sprues and runners are cut away and finishing takes place to create the finished product.

▲ A step-by-step guide to the main steps of sand casting and how it works

Moulding (plastics)

As well as metals, engineers often use plastics to make their products. Plastics have many beneficial properties and characteristics, which make them a very popular material to work with. Traditionally, plastics were produced from crude oil, a non-renewable resource, however **bioplastics** are produced from vegetable oils, corn starch and even recycled food waste. Plastics are also **self-finishing**. By applying different textures to the interior surface of the moulds used in shaping plastics, you can create any desired finish such as gloss, matte or textured. Following are some examples of the moulding processes for plastics.

Injection moulding

Injection moulding is a common moulding process that forces (injects) liquid plastic down a screw and into a mould. This process is accurate, good for high-volume production and creates little waste but is expensive to set up. Examples of products made in this way include PC cases and automotive parts.

Products that have been injection moulded still need some minor finishing when they emerge from the mould, such as trimming the sprue and any plastic 'bleed' from where the two halves of the mould meet.

> **Key Terms**
>
> **Pattern** A 3D copy of the item that is going to be cast. The sand is packed around the pattern and then the pattern is removed, creating a cavity that will be filled with the molten metal.
>
> **Expendable mould** A temporary mould that will be destroyed when the casting process is complete. It is not reusable.

WJEC Level 1/2 Vocational Award Engineering (Technical Award)

▲ Plastic keyboards can be made by injection moulding.

▲ Plastic bottles are made in the process of blow moulding.

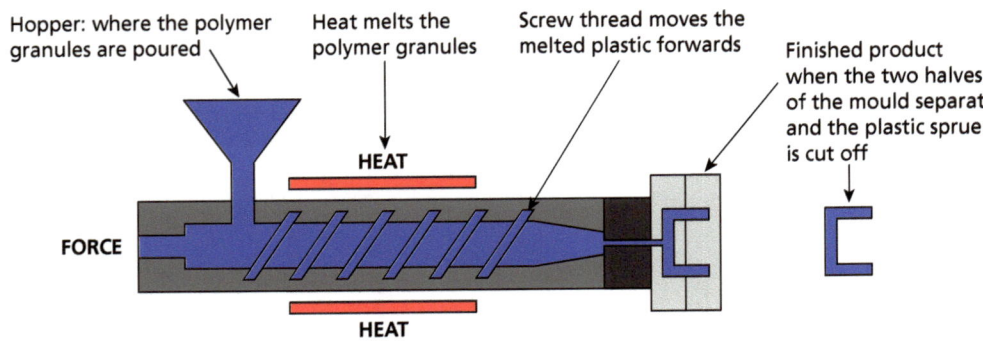

▲ Injection moulding

Blow moulding

In blow moulding, hot, malleable plastic is blown onto the outside of a mould. This process is inaccurate and low cost but is fast. Examples of products made in this way are drink bottles and cosmetics packaging. The four steps in the figure below show how a plastic product is created using the blow-moulding process.

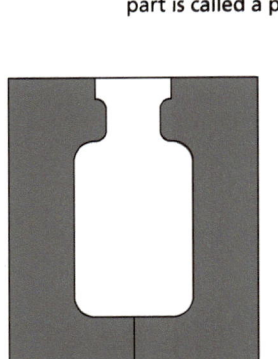

1 Mould

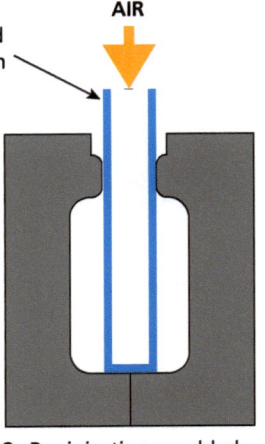

2 Pre-injection moulded plastic (parison) is clamped into the mould.

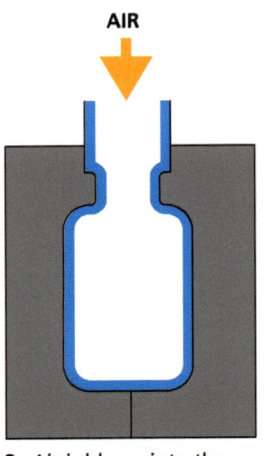

3 Air is blown into the parison allowing it to form to the outside of the mould.

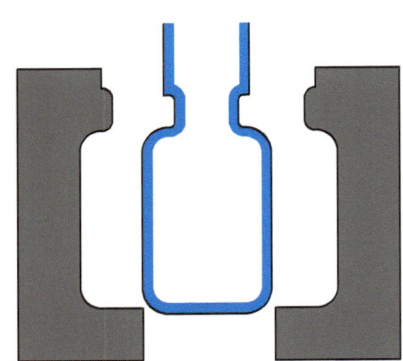

4 Mould separates leaving the finished product.

▲ Blow moulding

▲ Wheelie bins are made using rotational moulding.

▲ Vacuum forming is the process used to make safety hats.

Rotational moulding

Rotational moulding is another common moulding process that uses heat and gravity to allow plastic to form to the exterior of a mould. This process is inaccurate and low cost but is good for larger objects. Examples of products made in this way include grit bins, wheelie bins, bollards and street furniture. The four steps in the figure below show how a hollow plastic product is created using the rotational-moulding process.

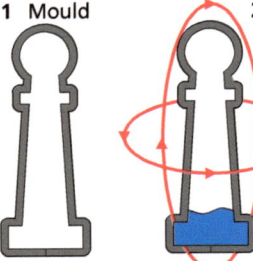

1 Mould
2 Polymer granules poured in, heat applied and the mould rotated.

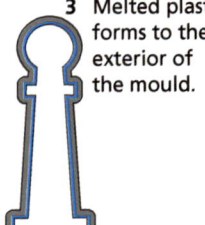

3 Melted plastic forms to the exterior of the mould.
4 Mould separates leaving the finished product.

▲ Rotational moulding

4 Implementing engineering processes

Formers

A former is a shape created by an engineer out of materials such as wood to create a template. These formers are created so that moulding processes, such as **vacuum forming**, can take place; the plastic forms around the pre-manufactured piece. Essentially, a former is a template to create a product.

Vacuum forming

In vacuum forming, a vacuum is used to form sheet plastic over a preformed mould. This process is inaccurate, low cost and used for low-volume production. Examples of products made in this way include chocolate packaging inserts, bike helmets and cutlery trays. The four steps in the following figure show how a plastic product can be created from a plastic sheet by the vacuum-forming process.

> **Key Term**
>
> **Platen** The part of the vacuum-forming machine that acts as a shelf. It has holes and can be raised or lowered as necessary.

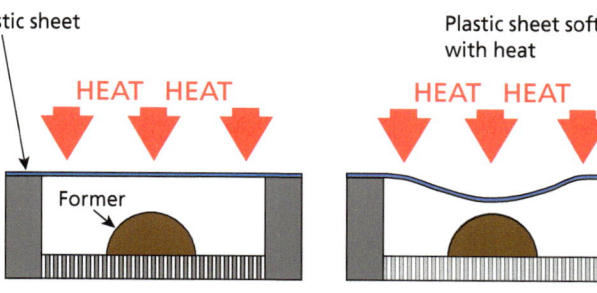
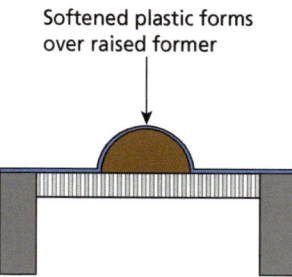
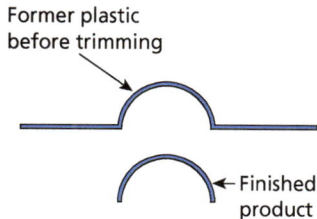

1 The former is placed on the **platen** and then lowered. A thin plastic sheet (e.g. high-density polyester) is then placed on the frame above it. Heat is then applied to the plastic sheet.

2 Heat continues to be applied to the plastic until it becomes malleable.

3 The platen is then raised into the plastic and a vacuum pump sucks out all the air, ensuring the plastic forms around the former.

4 The former is removed from the plastic and any excess plastic is trimmed away, leaving the finished product.

 *Vacuum forming*

Task 4.2

Identify the different engineering processes and their functions/uses.

> **Key Terms**
>
> **Aesthetic** The way something looks to the eye.
>
> **Corrosion** A gradual deterioration of metals caused by the action of air, moisture or a chemical reaction (such as an acid) on their surface. Corrosion can be described as the oxidisation of a metallic surface. When an iron surface corrodes, we say it has rusted.
>
> **Resistance** A material is said to be resistant if it resists a change.

Material finishes

Quite often, the materials that have been produced need to be **finished** as part of the processes used to create them. In other words, the products need to undergo another manufacturing process to give them a particular finish. Reasons for finishing a material can differ.

- A finish can be used for **aesthetic** reasons, to make the product look nice.
- The finish can be functional (for example, to give the product a rougher surface to provide grip).
- It can provide protection or **corrosion resistance**, to stop it rusting or discolouring.

Finishes for metals

Plastic dip coating: This finish is used mainly on steel. Metal is heated and dipped into plastic powder that melts and sticks onto the surface of the metal. It is good for anti-corrosion. A range of colours is available for aesthetic purposes.

Galvanising: This involves coating a ferrous metal (steel) with zinc (non-ferrous) to protect it against corrosion/rust. Galvanising creates a thin, non-ferrous layer of metal between the steel and the elements (rain/wind). It can be used for streetlights and fences as it has a very durable finish.

Anodising: Aluminium is placed in a bath of acid and an electric current is passed through it. Coloured dye is then added, which permeates the surface of the aluminium, adding colour. This process is not just used for aesthetic purposes, it also adds to the corrosion resistance of the aluminium part.

▲ A plastic dip-coated screw

▲ A galvanised steel fence

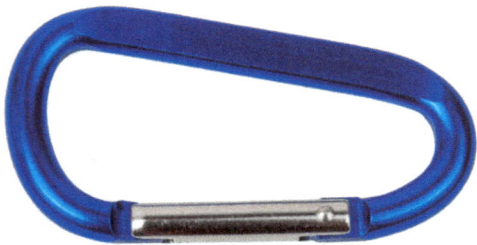
▲ An anodised aluminium carabiner

Powder coating: Similar to plastic dip coating but the plastic powder is sprayed on. This process is often used in industry, for example for coating white goods such as washing machines and dishwashers.

Blueing: In this process, steel is heated up and then dipped in oil. The oil permeates the surface of the steel to create an anti-corrosion layer that protects the steel against rust. The finish tends to be a blue/black colour. 'Cold blueing' can also be carried out using chemicals instead of heat and oil. The process is used for tools and by gunsmiths.

Painting: Paint creates a barrier between the metal surface and the elements for corrosion resistance. There are also aesthetic benefits from the many colour options. Metal is prepared first using a primer. Some paints are designed specifically for metals, for example Hammerite. The paint finish on products such as bridges, ships and goalposts, etc. needs to be maintained on a regular basis.

Enamelling: High temperatures are used to melt powdered glass onto a metallic surface to create a glass barrier between the metal and the elements. The benefits are corrosion resistance and aesthetic appeal, as lots of colour options are available. One of the most common enamelled items is the tin mug, but a lot of jewellery is also created using enamelling.

▲ Powder-coated panels on a white washing machine

▲ Enamelled mugs

Finishes for wood

Painting: Paint is one of the most common finishes for wood surfaces. The wood needs to be prepared first with a primer or undercoat before a finishing coat of paint is applied. The finish resists weather wear and many colour options are available for aesthetic appeal. Paint can be water based or oil based; it can be brushed or sprayed on.

Varnish: Tends to be applied when you want the grain of the wood to be seen (instead of covering it up with paint). Most varnishes are transparent, so you can still see the wood, and have a gloss or matte finish. Multiple layers can be applied to create a barrier between the elements and the wood itself. Polyurethane or 'yacht' varnish is very hardwearing and is used in the marine industry.

Stains: Wood stains tend to have a thinner, more watery consistency and are generally applied with a brush. Stains permeate the surface of the wood and protect against the elements. They also add different colour choices for aesthetic purposes.

Wax: Applying wax to wood creates a waterproof coating to the surface, which can be buffed/polished to create a very smooth, natural-looking finish. Wax is good for indoor products and is aesthetically pleasing if you want to see the natural tones of the wood itself.

▲ *Waxing wood*

Finishes for plastics

Plastic is a material with lots of great properties: strength-to-weight ratio, mouldable into any shape, various colours. It also has the wonderful property of being **self-finishing**. Unlike the other materials discussed, plastic does not have to go through another process to finish it. This can save companies money as they do not have to purchase extra equipment, invest in extra space, or train and pay more staff to finish their plastic products.

The finish on plastic products is chosen during the design stage and the interior of the plastic mould will have the chosen finish on it. So, when the plastic product comes out of the mould, it retains the surface finish of the mould.

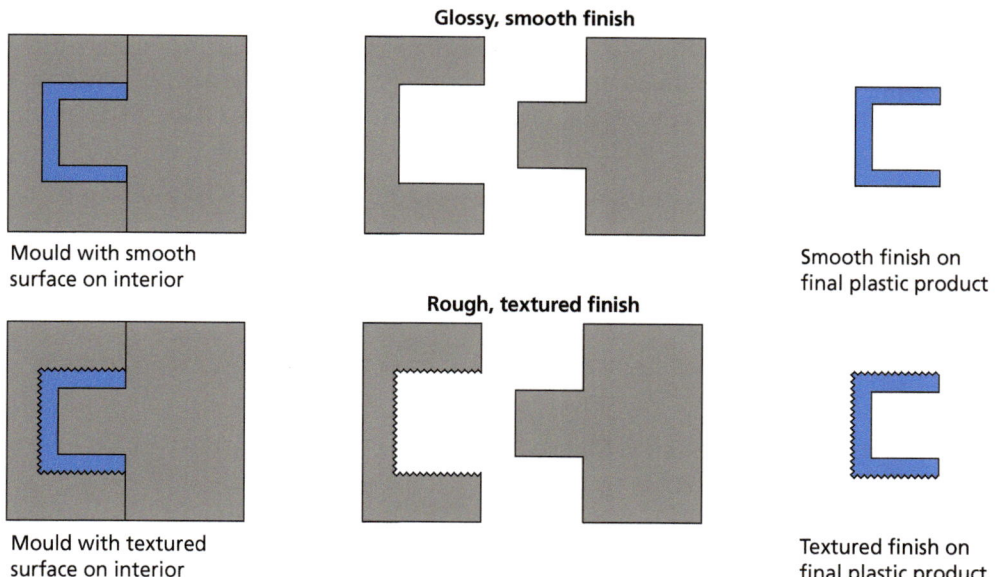

▲ *Example of the different finishes on a plastic product applied via a mould*

There are a handful of production processes, however, that do require the plastic products to be finished.
- **Cutting or sawing plastic:** This process could leave a rough edge with plastic burrs. The product may need smoothing out with fine-grade abrasive paper or even buffing on a polishing machine.
- **3D-printed products:** These plastic products tend to be made from plastic 'wire' that leaves ridges all around the product. They can be finished with fine abrasive paper.

▲ *A selection of plastic products manufactured by 3D printing*

Task 4.3

Identify the different engineering finishes and where they might be used.

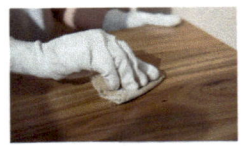

Evaluate the quality of engineered products

Once an engineer has designed and created a product to solve the problem that was provided to them in the brief, they need to assess the outcome against a range of criteria. The assessment may identify elements of the brief that the outcome does not satisfy, meaning that changes and improvements are required. Evaluating the outcome is also vital for the engineer, as they can identify any parts of the manufacturing process that could be improved, so that the product could be made more efficiently, or any elements of the product that could be improved, regardless of whether it was a successful outcome.

Against success criteria

At the start of any project, an engineer will be given a design brief by the client or customer. In the design brief, there will be key information provided to the engineer that the solution must have for it to be a successful product/outcome. The **success criteria** will be developed from this brief, as key information will be highlighted from the brief and interpreted by the engineer.

As we discussed in Chapter 2, engineers can measure the success criteria against the specification in the following ways:

- a **condensed brief** (a brief that is reduced in words to concentrate on the key areas)
- a **highlighted brief** (key areas of the brief are highlighted to stand out against the backdrop of words)
- a **bullet-point brief** (key areas of the brief have been bullet pointed, making it easier to read).

Examples of these can be seen in Chapter 6 (page 105), as this should be a task undertaken at the inception of the project.

Against design specification

Once an engineer has successfully completed a project and established a set of success criteria based on the client brief and thorough research, the next step involves evaluating the final outcome against the previously created **design specification**. The design specification serves as a detailed document outlining the engineer's expectations for the project's shape, colour, function, and other relevant factors. Essentially, it functions as a comprehensive list of success criteria that define the desired outcome.

During the process of devising a solution, whether on paper or as a prototype, engineers consistently measure and assess their progress against the established design specification. The following table outlines key considerations for writing and utilising a design specification in the evaluation phase.

Function	What does the product need to do?
Aesthetics	What should it look like?
Environment	How can the environment be protected in the manufacture of this product?
Materials	What materials should the product be made from?
Performance	Where and how will the product be used?
Target customer	Who is the target customer for this product and what are they like?

Design specifications will be covered in greater depth in Chapter 6.

Tolerance

Any technical drawing will include details of the tolerances that the product must stay within. For example, an engineer may be tasked with creating a small hand clamp which has multiple pieces. One of the main parts of this clamp is the central column. The technical drawing for the product states that the tolerance of this central column should be 100 mm +/− 1 mm. This allows the engineer to cut, file and produce a final central column with a length between 99 mm and 101 mm. Anything below 99 mm or above 101 mm would mean that part is not within the accepted tolerance.

Quality inspections

In the process of **quality inspection**, an engineer will check, measure or test the functionality or properties of their product. The engineer will then compare these results with the specific requirements that were set out in the client brief.

The three most common types of quality inspection are:
- pre-production inspections
- during-production inspections
- final inspections.

Pre-production inspection (PPI)

A pre-production inspection (PPI) is the evaluation process that an engineer would conduct before starting production of their product/project. This stage allows the engineer to identify quality risk before the manufacturing of the product begins.

During-production inspection (DPI)

A during-production inspection (DPI) is a carried out by the engineer while they are in the process of manufacturing their product or project. This is an important inspection technique as it will allow engineers to make modifications or improvements to the product before completion, therefore avoiding any future issues.

Final inspection

Final inspection, as the name suggests, is when an engineer conducts the last inspection of their product. At this stage, they will assess the outcome against the client brief and evaluate the project for any defects. Small improvements can be made if required.

Tolerancing 00 = ±0.2 00.0 = ±0.1 00.00 = ±0.05 Angular = ± 0°'30	Scale **1 : 1**	Size **A4**
All dimensions in mm		
3rd angle projection		
Do not scale		

▲ *All technical drawings provide information about tolerances.*

▲ Stages in the evaluation process

Evaluate own practices and processes

Not only must engineers consider how to evaluate the outcome of the project, but they must also self-evaluate their own performance. They will need to consider their strengths (what went well) and weaknesses (what they could improve on). This is vital for an engineer, as they will use this key information to improve their performance in future projects.

An engineer can evaluate using existing evaluation techniques. These techniques are recognised evaluation models and are also a useful tool in the engineering/design industry to evaluate projects, outcomes and processes. Such techniques include:
- total design model
- SWOT analysis
- listing advantages and disadvantages.

SWOT analysis

A **SWOT** analysis is a useful evaluation tool as it can help you to discover opportunities that can be exploited when working on a project, the strengths of your project and will also identify any weaknesses your ideas may have. SWOT can also discover any threats your project may have by looking at competitor products.

> **Key Term**
>
> **SWOT** The acronym SWOT stands for:
> - **strengths**: identifies good points about your project/idea
> - **weaknesses**: identifies areas of your project/idea that could make it fail
> - **opportunities**: identifies areas of your project/idea that you could exploit
> - **threats**: identifies areas that could cause trouble for the project/idea.

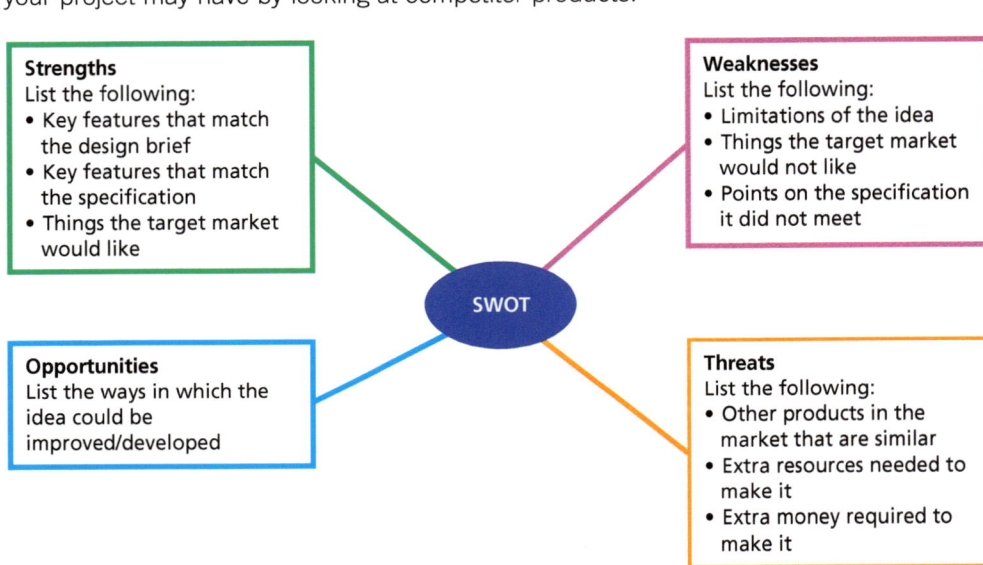

▲ A breakdown of the SWOT analysis method

Listing advantages and disadvantages

The final model of evaluating involves listing the advantages and disadvantages of your design ideas. This evaluation tool is probably the most common and one you should be familiar with.

By listing the advantages and disadvantages of your ideas against the success criteria, in a table similar to the one below, you are able to see which ideas are best suited to carry forward and develop further.

Success criteria	Advantages	Disadvantages
Design brief		
Specification		

Guide to coursework submission: Unit 1

5

This chapter looks at the first unit that you need to complete, with examples of tasks and pages that you can complete ready for submission.

Unit weighting

The WJEC Level 1/2 Vocational Award in Engineering is broken up into three different units. In each unit you will be asked to demonstrate your engineering skills and knowledge, and create evidence via portfolios, manufacturing skills and examinations. The evidence then needs to be submitted in an electronic format (your tutors will complete this part).

Each unit will have a different task and a different weighting:

Unit number	Evidence required	Weighting of unit	Time to complete (per student)
Unit 1	Manufacture a prototype Produce an electronic portfolio	40% (80 marks)	20 hours
Unit 2	Design/redesign part of prototype Produce an electronic portfolio	20% (40 marks)	10 hours
Unit 3	Complete a controlled examination	40% (80 marks)	1 hour 30 minutes

Examples for submission

Some of the difficulty in producing work for a technical subject such as engineering lies in understanding what you need to submit as evidence. The guidance offered by exam boards, although detailed, needs to be translated into actual tasks you can do.

The following examples are tasks that you could complete and that would satisfy the needs of the assessment grids (marking structure) for Unit 1.

IMPORTANT: The following are **EXAMPLES ONLY**. You must translate the needs and requirements of each unit with **YOUR OWN WORK**. You can use these examples as **GUIDANCE ONLY** and should not submit copies of the following work if you want to ensure you are not questioned about the originality of your submissions.

Unit 1: submission guide

In this guide you will find examples of evidence on what to create and submit for Unit 1 Manufacturing engineering products.

As noted above, this guide is made up of examples. You should interpret the needs of the specification yourself and produce your own work.

Assignment brief example

A large lighting company is looking to expand its portfolio of products to include a desktop lamp that incorporates aesthetically pleasing materials such as brass, bronze and aluminium and that can be polished to a high sheen. The lamp would be cordless and use battery-operated LED lights. This will ensure the lamp could be purchased by people going camping and therefore increase the potential customer base.

Your role as an engineer is to:
- translate the information in the engineering drawings
- plan and sequence each stage of production
- select the necessary tools, equipment and materials
- manufacture and assemble all the required parts
- evaluate the outcome and offer suggestions for further improvements.

Examples of drawings

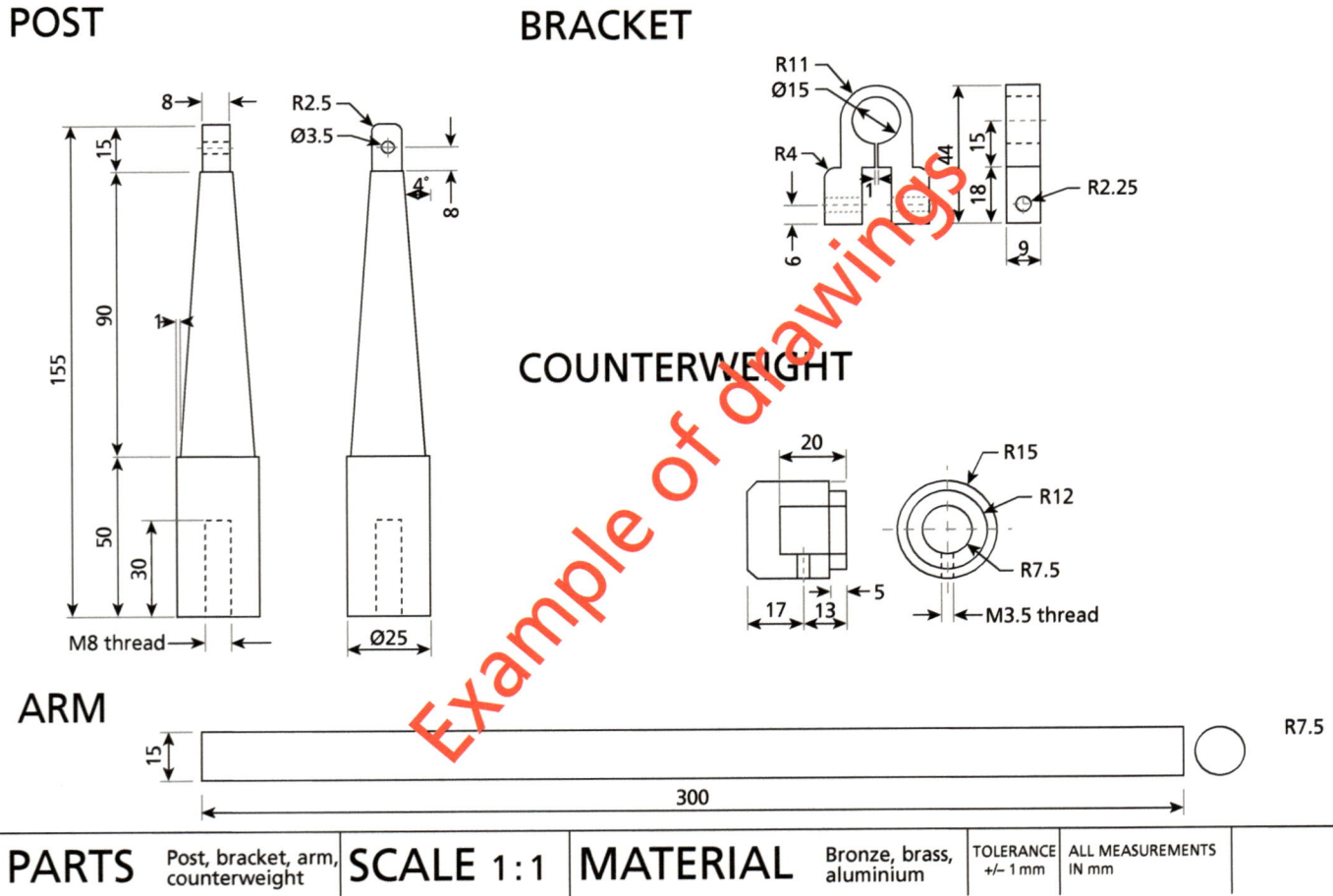

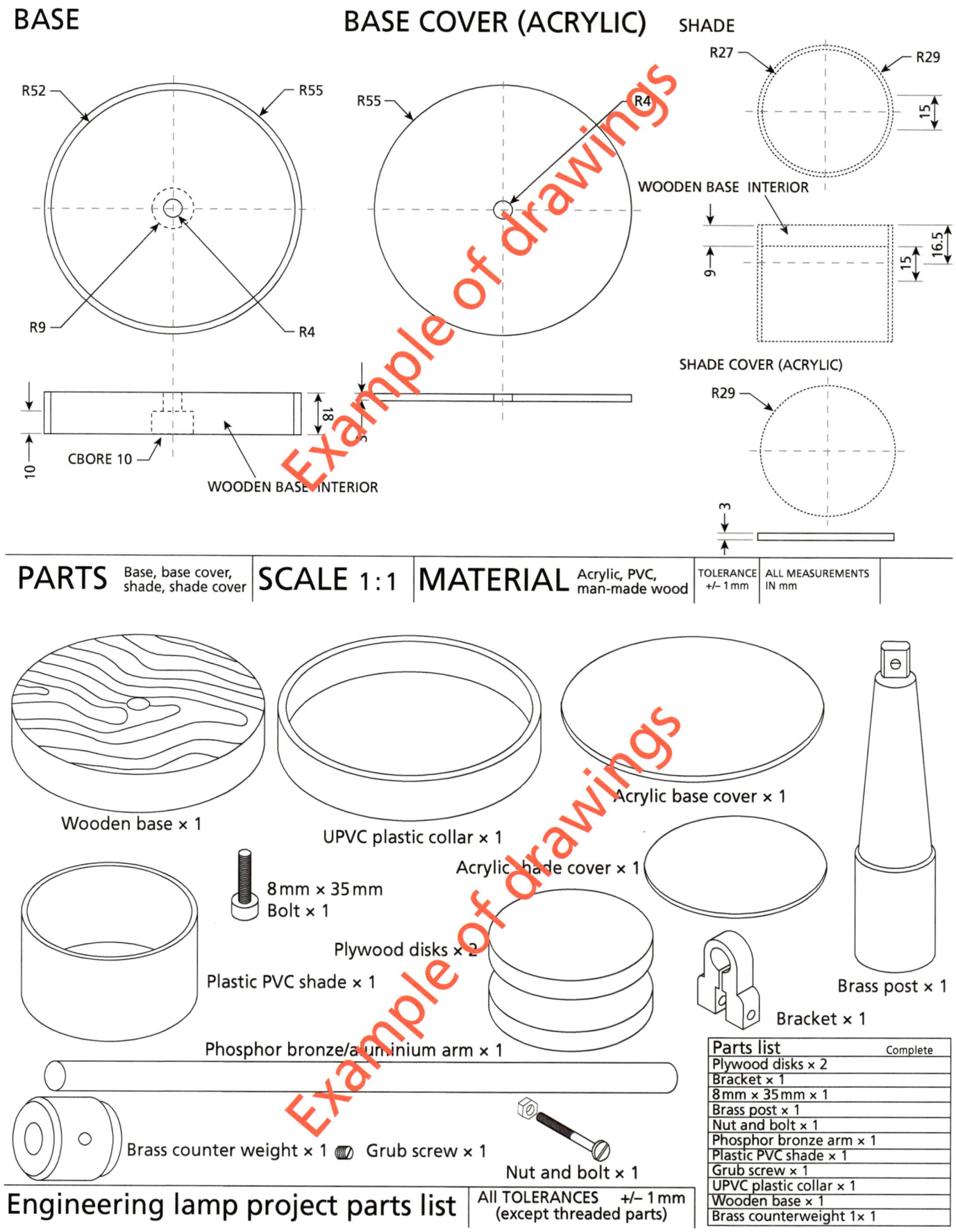

Requirements and marks available

Task 1(a)
Examine the provided engineering information to:
- identify the key parts and/or components to be produced
- analyse the required key information to produce the engineered product prototype.

[10 marks]

Task 1(b)
Collate the technical information needed to produce the engineered product in the workshop, including parts and/or components needed to complete the assembly to the given manufacturing specification.

[4 marks]

Task 2(a)
Select:
- suitable materials to produce the component parts from the engineering information, including identifying material stock and stock sizes
- necessary tools and equipment to produce the component parts from the engineering to begin production.

[10 marks]

Example page

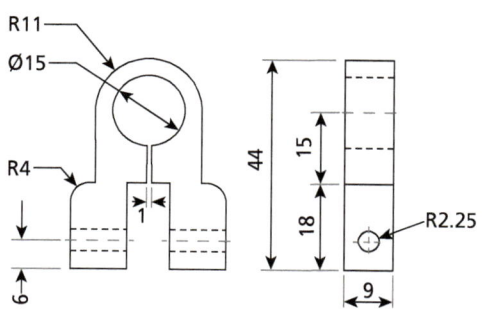

Tolerance	Jig needed?
± 0.3 mm	For polishing

Part: BRACKET			
Function	**Finishes**	**Needed properties**	**Possible materials**
• Attach arm and post • Create pivot point • Allow lamp's position to be repositioned	Polished finish	• Machinable • Low cost • Tensile strength	• Mild steel • Aluminium • Brass • Nylon • Acrylic
Brief requirements			
• The brief states that the product would need to incorporate aesthetically pleasing materials such as brass, bronze and aluminium. • The specification states that the product would need to be cordless and use battery-operated LED lights.			

Potential processes needed	Tools and equipment needed	Machinery needed	Machine speed
Marking-out	Engineer's blue Steel rule Vernier callipers Scriber		
Cutting	Hacksaw		
Filing	File		
Drilling		Pillar drill	360 RPM
Polishing		Buffer	3000 RPM
Material selection			
Options	**Properties**	**Choice**	**Justification/reason why**
Mild steel	Good tensile strength Low cost Relatively tough	NO	Difficult to knurl with school centre lathes
Aluminium	Easy to machine and shape Low cost Corrosion resistant	YES	Easy to finish with a knurl Easy to thread Accessible in workshop
Brass	Very easy to machine Corrosion resistant	NO	Expensive in relation to other metals
Final material stock-form and sizes		**Alternative components that could complete the task**	
Aluminium round bar Ø16 mm × 37 mm		Hex bolt Cap screw with knurled finish	
Potential problems			
1 Long queues for the centre lathe could limit access to machine. 2 Lots of processes to complete so could use a centre lathe for a long time. 3 Cutting an external thread with the die and die wrench could twist off the thin/turned down section as it is only aluminium.			

Task 2(b)

Using the provided engineering information, plan the stages of producing component parts.

Within this, you should:
- collate information into a time plan for the stages of production
- make contingency plans for potential unforeseen situations
- present detailed information in a way that would allow a third party to produce the engineered product.

[10 marks]

Example page

Production plan: TOP LOCKNUT							
Step/sequence	Process	Tools/equipment	Machinery	Machine speeds	PPE/Health and safety		Total time: 50 mins (5 minute intervals)
1	Marking-out	Engineer's blue Steel rule Vernier callipers Scriber	N/A	N/A	Wear apron		■□□□□□□□□□
2	Cutting	Hacksaw	N/A	N/A	Wear apron		□■□□□□□□□□
3	Drilling	N/A	Pillar drill	360 RPM	Machine guards Goggles Aprons Correct protocols		□□■□□□□□□□
4	Filing	Selection of files	N/A	N/A	Wear apron		□□□■■■□□□□
5	Wet and dry	Wet and dry paper	N/A	N/A	Wear apron		□□□□□□■■□□
6	Polishing	Buffing machine	Centre lathe	115 RPM	Machine guards Goggles Apron Correct protocols		□□□□□□□□□■

Contingency planning	
Problem	**Solution**
Queue for pillar drill	Begin other tasks/work on other projects
Pillar drill broken	Make part from other stock; use machinery in other schools/industries
School shutdown/illness/material not available	Create CAD model

Photographic evidence

Photo of your completed part	Photo of your completed part	Photo of your completed part	Photo of your completed part

Example page
Task 2(c)

Assess the potential risks for the main production stages involved in the production of the engineered prototype and recommend health and safety control measures to counter those risks.

[6 marks]

Part: TOP LOCKNUT	Risk assessing each production stage			
Manufacturing activity	Equipment	Potential risks	Control measures	Risk ratings Low/medium/high
Marking-out	Engineer's blue Steel rule Vernier callipers Scriber	Sharp tools (e.g. scriber) could cause scratches	Use tools correctly	Low
Cutting	Hacksaw	Sharp hacksaw blades could cut user	Hold work correctly Concentrate on task	Medium
Drilling	Pillar drill	Can trap loose articles of clothing in spinning chuck Can trap loose hair in chuck Can catch hands on chuck/workpiece Swarf flying off and into face/eyes Workpiece flying out of chuck	Follow protocols on correct machine use Use machine guards (if available) Tie back loose clothing and hair Wear correct PPE	High
Filing	Files	Loss of control could damage workpiece/fingers	Clamp workpiece securely	Low
Polishing	Buffing machine	Can trap loose articles of clothing in mop Can trap loose hair in mop Can catch hands on mop Workpiece flying off and across room	Follow protocols on correct machine use Use machine guards (if available) Tie back loose clothing and hair Wear correct PPE Manufacture and use jig	High
Parting off	Parting tool	Can trap loose articles of clothing in spinning chuck Can trap loose hair in chuck Can catch hands on chuck/workpiece Swarf flying off and into face/eyes Workpiece flying out of chuck	Follow protocols on correct machine use Use machine guards (if available) Tie back loose clothing and hair Wear correct PPE	High
External thread (M5)	M5 die and wrench (+ cutting compound)	Die could snap (high-carbon tool steel) Sharp edges could cut hands/arm	Hold workpiece securely (vice) Clear swarf by 'backing off' the cutting process	Low

Task 3

Produce an engineering outcome based on the details and data provided.

You must:
- use a range of engineering tools safely and effectively to produce the main parts and components of the engineered design prototype
- use a range of engineering equipment safely and effectively to produce the main parts and components of the engineered design prototype
- implement safe working practices and apply appropriate use of PPE during the entire production process.

[16 marks]

Task 4(a)

In the production of your engineering outcome, you must:
- apply skills in a range of engineering processes
- use a range of suitable materials.

[12 marks]

Task 4(b)

You must write a report that:
- evaluates the quality of the final prototype against the criteria given in the engineering drawings and specification
- evaluates your own practices and processes
- suggests improvements where appropriate.

[12 marks]

Example page

Evaluation of part: TOP LOCKNUT					
Function	Tolerance	Finish and quality	Personal performance	Improvements	Contingencies
The manufactured part functions well and meets the requirements of the specification and briefs	It has/hasn't met the tolerance limits of ± 1 mm	The part is finished with a high sheen polish	I worked well during the manufacture of this part All health and safety protocols were applied All machining protocols were applied Timings of each process were maintained	I could have taken the time to apply a better finish and remove all the scratch marks Drilling before filing would be a better option when sequencing the plan	No problems arose so no contingency plans were used
Photographic evidence					
	Photo of your completed part		Photo of your completed part		

All of the above could be completed for each part or you could amalgamate the required information for all parts together.

This next page will cover the whole unit requirement and lists any parts that need pre-manufactured or bought-in components.

Part	Bought-in components needed
Post	N/A
Bracket	N/A
Counterweight	N/A
Arm	N/A
Base	N/A
Base cover	3 mm acrylic; pre-manufactured (laser cut)
Shade	Plastic PVC piping 58 mm
Shade cover	3 mm acrylic; pre-manufactured (laser cut)
Bracket connections	M5 bolt and nyloc nut
Base and post connections	M8 hex bolt
Final assembly photographic evidence	

Photo of your assembled product

Photo of your assembled product

Total of 80 marks for Unit 1 (40% of course)

6 Designing engineering products

In this chapter you are going to:
- identify electrical components
- identify key mechanical components
- identify some key properties of component materials
- analyse a design brief
- identify key features of a design brief
- identify key features of existing products that relate to a design brief
- begin to develop solutions to a design brief using your learned communication skills
- learn how to analyse existing products effectively
- learn how key words can be used to identify and analyse products
- understand the term 'reverse engineering' and how to do it
- learn the differences between the component parts of a product
- understand how different parts of a product work together
- discover what a design specification is
- find out what information needs to be included in a design specification
- learn how to construct a design specification
- find out how to use prior learned knowledge to help write a design specification
- learn how to correctly identify metric measurements
- learn how to use simple mathematical formulae to solve engineering/mathematical problems
- learn how to work out the values of an electronic circuit
- learn how to apply simple ratios and equations to determine mechanical advantage in simple gears, levers and pulleys
- learn why we evaluate
- discover what an effective evaluation is
- find out how evaluations help us
- discover the different methods of evaluating
- learn how we can use different 'techniques' of evaluation.

This chapter will cover the following areas of the WJEC specification:

Unit 2 Designing engineering products: 2.1 Understanding function and meeting requirements		
• 2.1.1. Primary features of the given engineered product	• 2.1.2 Identifying features of other engineered products	• 2.1.3 Function of the proposed solution
Unit 2 Designing engineering products: 2.2 Proposing design solutions		
• 2.2.1 Generating a range of engineered solutions	• 2.2.2 Developing ideas through to a conclusion	• 2.2.3 Communicating design ideas
Unit 2 Designing engineering products: 2.3 Communicating an engineering design solution		
• 2.3.1 Producing an engineering specification	• 2.3.2 Drawing an engineering design solution that adheres to recognised standards	
Unit 2 Designing engineering products: 2.4 Solving applied engineering problems		
• 2.4.1 Using mathematical techniques for solving applied engineering problems		
Unit 3 Solving engineering problems: 3.4 Solving engineering problems		
• 3.4.1 Using mathematical techniques for solving engineering problems		

6 Designing engineering products

Primary features of the given engineered product

Before an engineer can decipher a client's brief and begin to work on their solution, they require a basic understanding of various electrical and mechanical components. Without this understanding, how would an engineer have the knowledge to create an effective solution?

As you may know, electronic circuits are populated with different **components**. These components all have different jobs to do and, when put together, can create electronic systems that perform specific tasks. An engineer needs to be able to identify electronic components to design circuits for electronic systems. When designing and drawing circuits, electronic components are represented by **symbols**. The following symbols are examples of the components that can be used in a range of circuits.

Component symbols are used as an easy way to communicate and visually represent a circuit. What if an engineer was in another country and there was a language barrier? Because the component symbols are used globally, the engineer in the other country would be able to understand the circuit through symbols and schematic diagrams.

Electrical components

Component	Description	Image	Component symbol
PCB	Printed circuits boards (PCBs) are the circuit boards used in most electronic devices. The boards acts as a medium on which electrical components can be connected. PCBs are typically made from an insulating material with layers of copper circuitry.		
Resistors	A resistor is an electrical component that limits or regulates the flow of current in a circuit. Resistors can also be used to provide a specific voltage for a device such as a transistor.	Band A – First digit; Band B – Second digit; Band C – Multiplier; Band D – Percentage tolerance	
Capacitors	A capacitor temporarily stores electrical energy by distributing charged particles on two plates. A capacitor can take a shorter time than a battery to charge up and it can release all the energy very quickly.		
Transistors	A transistor is a tiny semiconductor that controls the flow of current or voltage. In addition, it can generate and amplify electrical signals and act as a switch/gate for them.		
Diode	A diode is a device that allows current to flow easily in one direction, but severely restricts current flow in the opposite direction.		Anode (+) ▶▏ Cathode (−)

Component	Description	Image	Component symbol
Fuses	A fuse is a safety device in an electric plug or circuit. It contains a piece of wire which melts when there is a fault so that the flow of electricity stops.		
Variable resistor	Variable resistors are used to adjust the resistance in a circuit as a way of controlling the voltage or current.		
Thermistor	A thermistor is a component that can be used to measure the temperature of a device. It is a type of resistor in which the resistance is dependent on the temperature. In a thermistor, the resistance decreases as the temperature increases. In a temperature-controlled system, the thermistor is a small but important component.		
Cell	A single battery, like the type you put in a torch, is actually called a cell. It is only when you have two or more of these cells connected together that you call it a battery.		
Battery	A source of electric power consisting of two or more cells. It has external connections to power electrical devices.		
Switch	A switch is a simple device that is used to either break an electric circuit (open circuit – switch off) or complete it (closed circuit – switch on).		Open / Closed
Push-to-make switch	A push-to-make switch activates a circuit when a button is pressed. When the button is released the circuit is broken, and the flow of current is stopped.		
LED	A light-emitting diode (LED) is an electrical component which emits light when electricity passes through in one direction and prevents electricity from going in the reverse direction.		

6 Designing engineering products

Component	Description	Image	Component symbol
Bulb	A light bulb is a device that produces light from electricity. Light bulbs turn the electricity to light by passing current through a thin wire called a filament. The filament is usually made of tungsten, a material that emits light when electricity is passed through it.		
LDR	LDRs (light-dependent resistors) are used to detect light levels, e.g. in automatic security lights. Their resistance decreases as the light intensity increases. In the dark and at low light levels, the resistance of an LDR is high, and little current can flow through it.		
Speaker	A device that converts electrical energy into sound energy.		
Buzzer	A buzzer is a device that makes a sound when an external voltage is applied to it. It can only make one sound. The buzzer will get louder as the electrical current increases.		
Motor	An electrical motor is a device that converts electrical energy to mechanical energy. It works on the principle of the interactions between the magnetic fields of a permanent magnet and the field generated around a coil conducting electricity.		M
Voltmeter	A voltmeter is an instrument used for measuring the potential difference, or voltage, between two points in an electrical or electronic circuit. A voltmeter is connected in parallel with the component it is measuring.		V
Ammeter	An ammeter is a device used to measure current. To measure the current flowing through a component in a circuit, you must connect the ammeter in series with it.		A

The following figure is a simple circuit diagram for a circuit with four components.

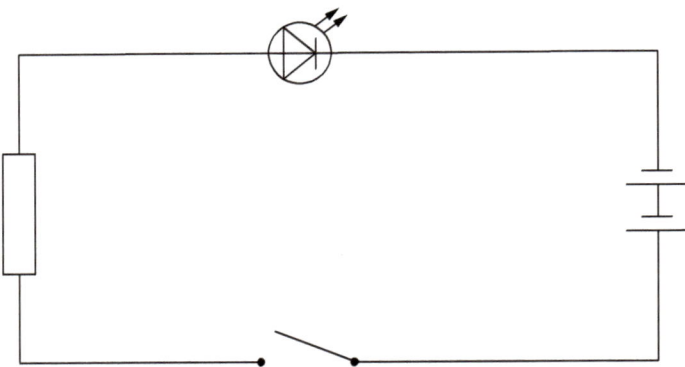

▲ A simple circuit diagram for a circuit that creates light with an LED.

By understanding the symbols used for components, you can identify the function of a circuit just by looking at the diagram.

Electrical values

The table below summarises the metric prefixes used, giving the prefix, the letter form and the multiplier each represents.

Letter	Word	Multiplier
p	pico	$\times 10^{-12}$
n	nano	$\times 10^{-9}$
µ	micro	$\times 10^{-6}$
m	milli	$\times 10^{-3}$
k	kilo	$\times 10^{3}$
M	mega	$\times 10^{6}$
G	giga	$\times 10^{9}$
T	tera	$\times 10^{12}$

Mechanical components: fixings

Nuts and bolts

Nuts are fasteners with threaded holes that secure to bolts, screws or studs and hold parts together. Once securely fastened, a nut prevents movement in connected materials.

Bolts are used to fasten and assemble parts through their aligned unthreaded holes, typically with the use of a matching nut. Bolts primarily consist of a shaft and a bolt head. The unthreaded portion of the shaft is called the shank, while the threaded portion is referred to as the bolt thread.

▲ A bolt is used with a matching nut.

Washers

▲ Washers come in various shapes.

Washers are machine components that are used with a screw fastener such as a nut and bolt. They usually serve to either keep the screw from loosening or to distribute the load from the nut or bolt head over a larger area. For load distribution, thin flat rings of soft steel are usual.

Screws

Screws are generally intended for use with a pre-drilled interior tapped hole or a nut. Machine screws are most often used for fastening metal parts securely together in various types of machinery or construction. Screws can come in a range of sizes and with various heads, depending on what the purpose of the job is.

Properties of component materials

Conductivity

Conductivity is a measure of the ease with which heat or electricity can pass through a material. A conductor is a material which gives very little resistance to the flow of electric current or **thermal** energy. An insulator is a material in which current and heat do not flow freely.

Friction

Friction is a force between two surfaces that are sliding, or trying to slide, across each other. For example, when you try to push a book along a carpeted floor, friction between the book and the carpet makes this difficult. Friction always works in the direction opposite to the direction in which the object is moving, or trying to move.

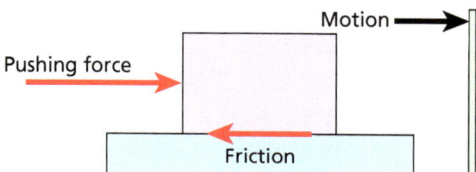

▲ Friction is a force between two touching objects that works against movement.

Durability

Durability is the ability of a material to last a long time without significant deterioration. A durable material helps the environment by conserving resources and reducing waste and avoiding the environmental impacts of repair and replacement.

Quality

In engineering terms, quality is the standard by which an engineer measures and improves their product during the development process. The quality of the product can be defined by multiple different criteria, including reliability, aesthetics and performance.

▲ Different types of screw

> **Key Term**
>
> **Conductivity** A measure of the ease with which heat or electricity can pass through a material.
>
> **Thermal** Relating to or caused by heat.

Design briefs

Introduction

We have already learned how engineers need to communicate effectively with both non-engineers and other engineers. It is essential that an engineer can communicate with customers/clients and with other colleagues within their company. These groups of people would generally be the ones who give engineers the tasks they need to perform. As we learnt in Chapter 2, the formal name of these given tasks is a **design brief**. This document of instructions from the client was introduced as part of Unit 1 Manufacturing engineering products.

A design brief can come in many forms. It could be a casual discussion between the client and engineer on what is needed or it could be a formal written document, outlining lots of specific areas that need to be addressed. Either way, it is still the engineer's job to be able to analyse a brief effectively, identify the most important features of the brief and then propose a number of solutions that would satisfy the brief.

Following is an example of a written design brief. This could be a typical series of statements that an engineer would receive from clients/colleagues.

Example of a design brief

A large manufacturer of bicycles and bicycle add-ons is looking to produce a new type of cycle specially to be used for shoppers who are high-street shopping. The solution could be a bicycle, tricycle or any cycle combination that would be best for the shopping task. The cycle solution would need to be made from existing materials that the manufacturer currently uses, to ensure that production of the new product is feasible.

There must be sufficient space to transport shopping items (clothing, groceries, etc.) as well as an option for security. The new cycle solution would need to be used by a range of people with different heights and sizes. The target market would also need to be able to afford to purchase the new cycle, so the cost of manufacture would need to be considered.

In this brief there are key features that an engineer would need to produce successful solutions. However, there are also pieces of information that may not be that useful when developing solutions. So, let's highlight all the key features that can be seen.

Design brief with highlighted key features

A large manufacturer of bicycles and bicycle add-ons is looking to produce a new *type of cycle* specially to be used *for shoppers* who are high-street shopping. The solution could be a bicycle, tricycle or *any cycle combination* that would be best for the shopping task. The cycle solution would need to be made from *existing materials* that the manufacturer currently uses, to ensure that *production of the new product is feasible*.

There must be *sufficient space to transport shopping items* (clothing, groceries, etc.) as well as an *option for security*. The new cycle solution would need to be used by a *range of people with different heights and sizes*. The target market would also need to be able to afford to purchase the new cycle, so the *cost of manufacture* would need to be considered.

Once the identifiable key features have been highlighted, you can rewrite the design brief in a **condensed** format, or even write a **bullet-pointed list** of things that the client/colleague has asked you to produce.

Condensed brief

Develop a new cycle for shoppers. It can have any cycle combination, made from current-use materials and it should be easy to manufacture. It needs storage space, needs to be adjustable, lockable and has to be low cost.

And now let's try a very simple bullet-pointed list of key features.

Bullet-point brief list

- Cycle for shoppers
- Storage
- Two, three or four wheels
- Adjustable
- Existing materials (manufacturer)
- Possibly lockable
- Easy to make (manufacturer)
- Low cost

Once the list has been compiled, you can then look at prioritising it, so the most important tasks/areas to look at can be dealt with first. This also helps to identify what resources might be needed and when to obtain those resources.

Prioritised bullet-point brief list

Order of importance:

1 = most important

8 = least important

1. Cycle for shoppers
2. Low cost
3. Storage
4. Adjustable
5. Existing materials (manufacturer)
6. Easy to make (manufacturer)
7. Two, three or four wheels
8. Possibly lockable

As you can see, having the knowledge and skill to extract the key features of a design brief can benefit an engineer when trying to identify exactly what is needed in an engineered solution.

> **Top tip**
>
> Having an easily identifiable list of key features can save time and reduce confusion. It also allows the engineer to prioritise tasks and resources.

Task 6.1

Read the following brief. Using your new knowledge of how to analyse design briefs, pick out the key features of the brief and then, in your notebook, write a shorter, more concise version. Once you have completed that exercise, write a prioritised list of the key points with the 'most important' first and the 'least important' last.

Compare your prioritised list to the original brief. Which would be easier to work with and why?

Design brief

A large car manufacturer is looking to develop a 'universal' steering wheel system for its new range of cars. The new 'customisable' car industry has proven to be a success and the manufacturer wants to be part of the growing trend.

It is exploring the possibility of each customer having their own bespoke, individually designed steering wheel that they can quickly unclip and reuse in their other car models if needed.

The new steering wheel must be fully customisable, with the option to swap out designs, styles and accessories, with a view towards developing more options in the future. The new steering wheel must have a universal fitting system that can be used on all other car models from the same manufacturer. The new wheel must be easy to attach to and remove from the steering column with no more than one button or lever to perform this function.

You must consider how much the universal steering wheel would cost as well as the potential for improved security.

Creating solutions from a design brief

Once you have developed the ability to identify and extract the key features of a brief, you can use them to start creating solutions that go towards satisfying it.

> ### Task 6.2
> This task consists of two different design briefs.
> - Identify the key features of each brief.
> - Draw a solution using a recognised drawing technique.
> - Annotate and describe the key features (including material choice and explanation of why the material was chosen).
>
> ### Design brief 1
> A large furniture manufacturer has asked you to design a seating solution for pupils in schools. The solution should fit all pupils in a secondary school. It must be comfortable to sit on and be extremely durable. The school's contract for this new seating solution does not come with a great deal of money, so your finances are limited. The seating solution will be placed in all new schools. Note that all new schools have to conform to a green/environmental policy.
>
> ### Design brief 2
> Manchester United has asked you to come up with a solution to the problem of where to seat their coaches at their training grounds. The club wants somewhere for the coaches to sit when watching the team train. The solution will need to be movable. United wants the coaches to stay dry and be able to have an unrestricted view of the players. The solution will be placed on the training grounds and money is no object.

> **Top tip**
> When creating designs, use your drawing skills and stick to drawing in standardised forms, such as isometric.

> **Top tip**
> The design annotations should refer to the brief and explain how the key features are represented in the design.

> **Key Term**
> **Research** The process of finding things out.
> **Aesthetics** How a product or item may look. Is it aesthetically pleasing (nice to look at)?

Identifying features of other engineered products

Introduction
Engineers tend to be creative and innovative problem solvers. They use their skills and knowledge to specify materials and develop designs, and are key workers when creating new products. However, engineers also develop skills that enable them to learn from past and existing products. By learning from other solutions, engineers can improve their knowledge base and skills that will, in turn, enable them to optimise their engineering solutions and create better, more efficient products in the future. This process of 'finding things out' is also known as research.

Conducting research
When an engineer conducts research, they need to find out the key elements of existing products. The following section details some of the elements of a product that can be researched.

Aesthetics
What does the product look like? An engineer should think about the colour, finish and texture and consider what may have inspired its design. The engineer would be able to provide an opinion on the appearance of the product.

▲ Carrying out research is an important part of product design.

6 Designing engineering products

User/customer/client needs
Customer needs refer to the things that customers look for or require when purchasing a product or service from a business. When an engineer understands their customers' needs, they can tailor their products and marketing plan to better serve them.

Key customer needs an engineer must consider are:
- a fair price
- a good service
- a good product
- to feel valued.

Safety
Engineers need to consider what safety considerations have been made in the design of the product? Does the product consider the safety of the user? Why is the safety of the product important?

Ergonomics
Ergonomics is the science of designing products, systems and environments for human use. This means applying the characteristics of human users to the design of a product; in other words, matching the product to the user.

In order to achieve this, data about the size and shape of the human body is required. This branch of ergonomics is called anthropometrics.

Anthropometrics
Anthropometrics deals with human measurements, in particular their shape and size. For many products, systems and environments, complex data is required about any number of critical dimensions relating to the user, such as height, width or length of reach when standing. Therefore, anthropometric data must consider the greatest possible number of users.

According to the principles of anthropometrics, a designer would ignore the smallest and tallest users and design a product to fit the remaining 90%, who account for the greatest number of users.

Mechanisms
In mechanical engineering, a **mechanism** is a device that converts input motion and force/torque to output motions and force/torque. The four basic types of mechanism movements that an engineer might analyse in an existing product are:
- **Linear motion** – a product is exhibiting linear motion if it moves in a straight line, such as a train moving down a track.
- **Rotary motion** – movement around an axis or pivot point, for example a wheel.
- **Reciprocating motion** – involves a repeated up-and-down or back-and-forth motion, as shown by a piston or pump.
- **Oscillating motion** – a curved backwards and forwards movement that swings on an axis or pivot point, like the movement of a playground swing or a clock pendulum.

Electronics
Electronic products are any manufactured product or device, or component part of such a product, that has an electronic circuit. An engineer may study an existing product to determine the functioning of its electronic components. Key electrical components have been noted on page 101.

Key Term

Ergonomics An applied science concerned with designing and arranging things people use so that the people and things interact efficiently and safely.

Anthropometrics The science that defines physical measures of a person's size, form and functional capacities.

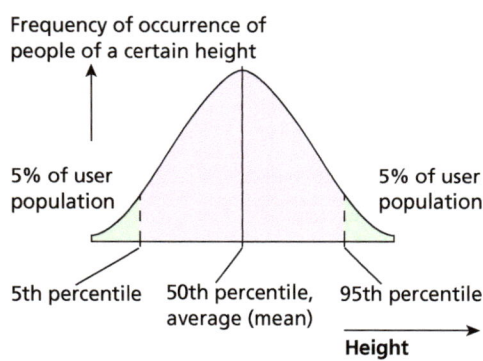

▲ Products tend to be designed for the 90% of people who fit between the extremes.

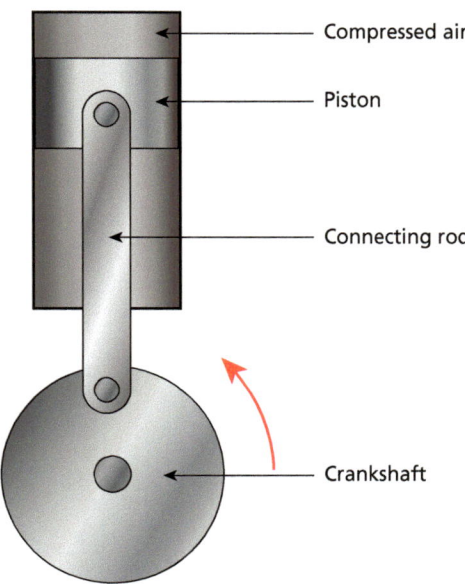

▲ This mechanism would demonstrate reciprocating motion.

109

Sustainability

Engineers will also consider **sustainable** engineering. This is the process of designing or operating systems such that they use energy and resources sustainably, in other words, at a rate that does not compromise the natural environment, or the ability of future generations to meet their own needs.

Research techniques

There are several ways of learning from/researching existing products and here you will learn about two of the most common techniques:

- **product analysis**
- **reverse engineering**.

Product analysis

Product analysis is a way of analysing products against specific **criteria**. So, you could look at and feel an existing product (such as the mobile phone in your pocket) and then ask: Does it look good? Is it easy to hold? Is it easy to use? Is it heavy or light? By asking these questions, you begin to find answers that would help you create a mobile phone with all the best features from the one you have analysed.

What other questions should you ask when analysing existing products?

Fortunately, there exists a model you can use that contains many of the headings and titles that could be used to analyse a product. However, be aware that this is a simple model and only has the most basic titles. Adding more titles (asking more questions) is good practice as it allows you to gather even more detailed information. This model, **ACCESS FM**, is detailed below.

> **Key Term**
>
> **Sustainable** A product is sustainable if its manufacture can be maintained (for example, if it is made from renewable resources).
>
> **Product analysis** Looking at, feeling and maybe using a product to see how it works.
>
> **Criteria** Specific headings or titles.

Aesthetic
- The way the product looks
- Does the target market think it looks good?
- Has it been finished well with appropriate colours?

Cost
- How much does the product cost to buy?
- How much to make?
- How much to run?
- Can the target market afford it?
- Is it a premium or low-cost product?

Customer
- Who is the customer/target market?
- What appeals to the customer/target market and why?
- Do the customers/target market need or want the product?

Environment
- Where will it be used?
- How will it be used there?
- What resources were used to produce it?
- Is it a sustainable product?
- Can it be recycled?
- Does it have a negative or positive impact on the local and larger environment?

Safety
- Is the product safe to use for the customer/target market or in its environment?
- Does it meet all safety regulations?
- Is it safe to make?

Size
- What numbers or values can you attribute to the product?
- Why does it need to be that big/small?
- What sizes would be applicable to the customer/target market?
- Can the product be scaled (up or down) to make it better?

Function
- Does the product function/work well?
- Does it perform the job it was meant to?
- Does it work for the customer/target market?
- Does it function well in its environment?

Materials and manufacturing techniques
- What materials have been chosen to make the product and why?
- What properties are needed from the materials to allow the product to function well?
- Are the materials renewable/non-renewable?
- Are they recyclable?
- What processes were used to make the product?
- Are the processes suitable or can you specify better suited ways of making the product?

▲ *ACCESS FM*

> **Top tip**
>
> As mentioned ACCESS FM is a good starting model to use when asking questions. However, for greater detail you may want to analyse a product using additional titles.

If ACCESS FM does not have the title you need to be able to find something out when analysing a product, you may need to add your own specific titles and headings. The following table contains some other headings and titles you may wish to consider.

Further headings	
Anthropometrics (from the Greek *anthropos* = human and *metron* = measurements)	• What size is your target market and how would that affect the sizes and positions of your product (think of a child's car seat)? • There is a lot of anthropometric data accessible on the internet that would be useful if you were creating furniture for children.
Ergonomics (similar to 'Function' in ACCESS FM)	• Does your product function well in relation to the customer/target market? • Will the sizes, positions, textures, finishes, material choice be affected by who you are designing for? • Does the product function in its environment? Is it easy/comfortable to use?
User requirements	• Does the product meet all the needs of the potential users? • Is there anything the product does not do well for the users?
Power/energy use	• How is the product powered? • Is it powered by battery, mains or solar? • Does it need a lead? • Is it portable power? • Can it be recharged?
Production levels	• How was the product produced? • Is it quick and one of millions in a factory or did a craftsperson make it over a longer period? • How will this affect cost etc.?
Legal requirements	• Is this product a copy or being copied? • Are there any innovations you can or cannot use?
Limitations/constraints	• What constrains the product? • Are there limits to its function? • Is the product limited to an environment/customer/time to use?

Key Term

Innovation From the Latin *innovare*, meaning to make new. To take something that already exists and improve it.

Reverse engineering

Reverse engineering is a very similar process to product analysis, as both methods involve the analysis of existing products to gain information that could be used to improve engineering solutions. As with product analysis, you can use headings and titles to analyse the product. However, instead of just analysing the products, in reverse engineering you dismantle a product piece by piece to discover things such as:
- how it was put together
- how it was manufactured
- how all the component parts interrelate
- where and what component parts are hidden
- where the hidden, innovative parts are that could be vital for the product to function.

Engineers often approach the reverse engineering task from two different perspectives:
- **external analysis**
- **internal analysis**.

The external analysis of a product can focus on (but is not restricted to) the information the product gives to the user such as look, feel, interactive areas (buttons/levers, etc.), properties and materials.

The internal analysis of a product can focus on (but is not restricted to) the manufacturing processes, assembly, hidden components, fixings, maintenance, materials, properties and how the components interrelate with each other.

Following are some headings and titles you can use to analyse an existing product when using reverse engineering:

External:
- ACCESS FM (plus others)

Internal:
- ACCESS FM (plus others)
- components
- assembly
- repair/maintenance
- recycling
- interrelation
- innovation

▲ *Recycling is an important consideration when analysing products.*

Identifying the component parts of a product

When analysing existing products, it is good practice for an engineer to be able to categorise the various components that are used in products. By categorising the components, you can quickly identify different systems that go towards creating the product and then see how those systems interrelate and how they work together to make the product function.

The different categories of components that you need to look at identifying are:
- component parts
- electrical components
- mechanical components
- materials and properties.

Task 6.3

Following are two images of a simple hairdryer. On the left is an internal view of the product and on the right is a view with the case/cover fitted.

Here is the list of component parts:
- case
- element cover
- dust cover
- heating element
- insulated copper wire
- screws
- wire flex
- speed switch
- heat switch
- fan
- electric motor.

Look at the following categories of components in the hairdryer:
- component parts
- electrical components
- mechanical components
- materials and properties

Sketch the hairdryers in your notebook and, using arrows, identify and label all the relevant parts for each category. Note that some parts of the hairdryer will fit into more than one category.

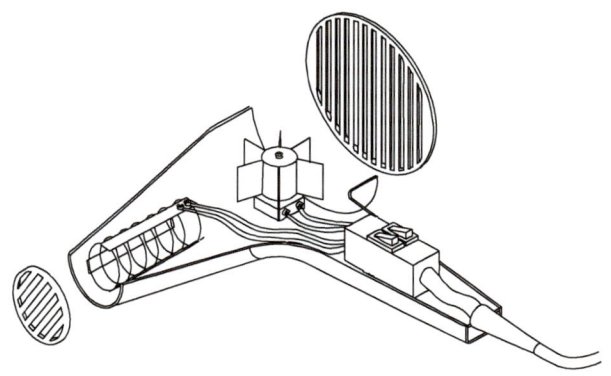
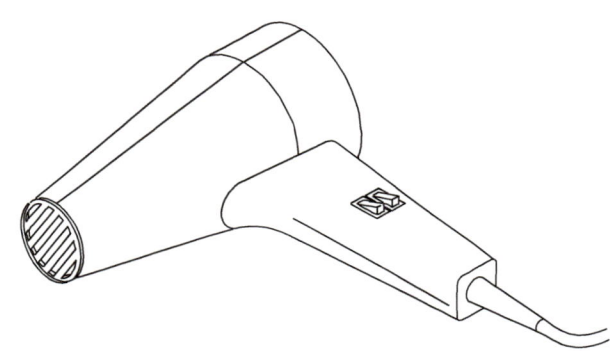

Task 6.4

Take a picture of a product from home or the internet and attempt to identify and label the:
- component parts
- electrical components
- mechanical components
- materials and properties.

Present your work on an A3 sheet. You can do this by hand, on a computer or use both methods.

How products function

Engineers need an understanding of how products function. Existing products tend to be made from lots of smaller component parts that all work together to create a functioning product. Sometimes the product will have moving parts/mechanisms and sometimes the product could be stationary but still have several different components, including a finish (see page 85 of Chapter 4 for more on finishes).

You have already learnt how to identify the different component parts of a product and can label individual components in isolation. The next stage is to see how those components work together to create a functioning product and explain how the component parts interrelate to make the product work.

WJEC Level 1/2 Vocational Award Engineering (Technical Award)

Seat for user to sit on. Connected to the frame (the strongest part of the bike).

Frame to hold the weight of the rider and support all the needed components of the bike.

Handlebars linked to forks and front wheel. Enable rider to turn the front wheel and steer the bike. Also host brake levers and gear-changing levers.

Rear wheel to create forward motion.

Brakes that use friction on the wheel rim to slow down the rotation of the wheel.

Cassette of gear cogs and freewheel, to increase speed by selecting different cogs, which is connected to the crank, pedals and gear selector lever on the handlebars.

Pedals and crank are connected to rear wheel by way of a cassette and chain. Cranking (turning) the pedals creates a rotational motion with the rear wheel, thereby creating forward motion for the bike.

Spokes to distribute weight evenly through the wheel rims.

▲ *All the component parts of this bicycle have been labelled with explanations of their function.*

Task 6.5

1. From the list below, identify the component parts of the drill shown in the top figure.
2. Copy the image of the drill into your notebook and using arrows, where needed, draw links between the component parts that work directly together.
3. Write down and describe how the component parts interrelate to create a functioning drill.

Component parts:
- battery pack
- variable resistor
- trigger
- electric motor
- gearing
- chuck
- drill bit
- copper wires
- wire connectors

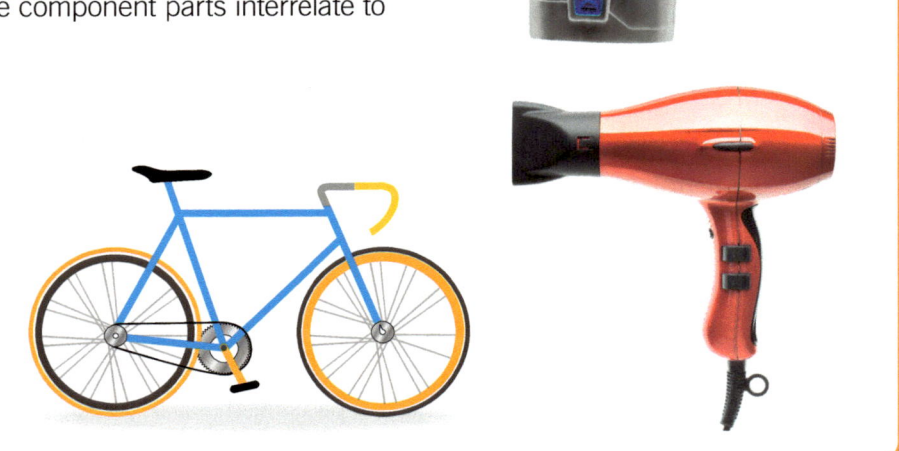

6 Designing engineering products

> **Task 6.6**
>
> On a separate A3 sheet of paper, select a product (drill, hairdryer, bicycle or another of your choice) to analyse.
>
> Place a picture of your chosen product in the middle of your A3 page and draw links between the main component parts.
>
> Describe how the component parts interrelate to perform a function.
>
> Try to choose a product/picture where you can see the component parts.
>
> Try to choose a simple product.
>
> Use lots of annotation to explain/describe the functions.

▲ *The disassembled components of a mobile phone*

Producing a design specification

Introduction
As you are aware, engineers are often asked to create new solutions to existing problems. When most of the research work has been completed (analysing briefs, product analysis, reverse engineering, etc.), engineers create a written document listing all the things the new solution will or may need in order to be successful. This list of success criteria is known as a **design specification**.

Design specifications should only be written once relevant information has been gathered from the research and can be used to define the criteria for success. Once the design specification has been written, it can then be used as a guide for the engineer to check progress against, as well as measure the developments and solutions against. The design specification is also a great tool to use when evaluating the success of your final outcome.

What do you need to include in a design specification?
Much like the product analysis process, a design specification needs to contain lots of different criteria that are relevant to the brief and the solution you are developing. These criteria can be written in headings and titles, such as what materials would be used, what the product will look like, sizes and other appropriate criteria.

> **Top tip**
>
> ACCESS FM could be a good starting point for design specification criteria.

How should the criteria be written down?
Once you have completed the research, you will understand what the solution must be or must have. You can then begin to write statements or points, explaining your solution. A typical design specification point could be written like this:

The solution must be made from a non-ferrous metal to ensure it will not corrode.

This specification point would likely be under a heading of 'materials'.

What else makes a good design specification?

A hierarchy
A design specification should contain a **hierarchy**. This is essentially a list where items are ranked from the most important to the least important. By using a hierarchy in a design specification, an engineer can identify which points should be concentrated on first. The available resources can then be targeted at these things in the first instance.

115

Key
1 = most important
10 = least important

Key
★★★★★ = most important
★ = least important

Ways of splitting your specification up into a hierarchy
You could split all the specification points into two lists:
- a list of **essential** specification points
- a list of **desirable** specification points.

Of these, the list of essential points is the most important.

You could also assign criteria or numbers against each specification point, with a key explaining what each of the numbers or symbols mean.

Qualitative and quantitative data
A good design specification should contain **qualitative data** and **quantitative data**. These are sets of data (information) that complement each other when either gathering or using information. In a design specification, both sets of data can be used as guides for success criteria and as areas that can be used to measure your solution against.

The difference between qualitative and quantitative data
Qualitative data relates to:
- things that may not be strictly measurable
- descriptions, feelings, opinions and emotional responses.

Quantitative data relates to:
- things that can be measured and weighed
- facts and numbers.

The following figure illustrates some questions relating to qualitative and quantitative data for a typical mobile phone.

Mobile phone

Qualitative data:
- How does it look?
- How does it feel?
- Do you have an emotional response?
- How do you feel when using it?

Quantitative data:
- What is it made from?
- How big is it?
- How heavy is it?
- How much does it cost?
- What is the scale of manufacture?

▲ *Quantitative data is about things that can be measured or weighed.*

> **Top tip**
> Your measurable criteria don't just have to be numbers, such as sizes. They could also include colours or even your end users' opinions.

Measurable criteria
A good specification should also contain points that are clearly measurable. This will allow you to evaluate your solution against your written success criteria (specification).

Task 6.7
Read the following detailed design brief and then copy and complete the blank specification into your notebook.
- Remember to check the design brief for any key features.
- A specification template has been created for you to copy but you could create your own style of specification and include any of the criteria you have learned about so far (hierarchy, qualitative/quantitative data, etc.).
- One essential criterion has been written as an example.

6 Designing engineering products

Design brief

You have been asked by an engineering automotive company (e.g. Ford or Toyota) to develop a solution to the following problem:

With the advances in tyre technology, punctures are becoming less of a problem to the daily motorist. However, it is a problem that cannot be completely eradicated and motorists are still having to pull over on busy motorways and roads to change wheels. This is a time-consuming process and a certain amount of force is needed to lift the wheel in place (onto the bolts). We would like you to develop a solution to make the lifting and placing of a spare wheel onto the raised car easier for motorists. It must be:

- small enough to fit in the boot of a car and not take up any extra space
- low cost enough that it can be made cheaply and in the millions
- strong enough to support a spare wheel
- simple to use.

▲ Changing a tyre can be difficult.

Specification   Engineering design .... Name ....

Essential criteria   **Desirable criteria**

Materials
It should be made from a metal with good tensile strength
It should be made from a metal that is easy to fabricate
The materials should be finished for anti-corrosion purposes

▲ A sample design specification for your solution.

Solving applied engineering problems

Understanding how to use mathematical techniques is a fundamental skill that all engineers should have. Engineers are commonly faced with problems that lead to the questions: how big, how heavy or how much? When building bridges, engineers need to know how heavy the traffic might be, which would be a deciding factor in what materials they would specify for the task. On a smaller scale, an engineer would need to know what components would be used in a circuit and therefore how much power would be needed to run the circuit.

Mathematics and engineering go hand in hand and an engineer would not progress very far without some knowledge of this subject. Mathematics can sometimes be seen as a difficult subject and might deter some people from considering engineering. However, if you approach mathematics in a simple step-by step way, it can be easy to understand. In this section you are going to learn some techniques in easy-to-access, step-by-step processes that will give you a greater understanding of how mathematics can be applied to simple engineering problems.

 **Key Terms**

Measuring To discover the exact size, amount, etc., of something, or to be of a particular size.

Measurement The size, length or amount of something, as established by measuring.

Measuring

Engineers need to have a basic understanding of how to **measure** to ensure their completed project is accurate. Having this understanding of **measurements** is also an important skill when preparing and ordering materials.

The metric system

▲ This ruler shows lengths using the metric system.

The metric system is a system used for measuring things. It is used worldwide in calculations and research. The metric system is used to measure length, weight and capacity and is what engineers will use when measuring materials in the workshop. Engineers should not confuse the metric system with the imperial system that is still used in America. On measuring equipment, such as steel rules, you will often find the metric system on the top and the imperial system on the bottom.

The following table summarises some units of length, weight and volume from the metric system.

Length	Weight	Volume
millimetre (mm)	milligram (mg)	millilitre (ml)
centimetre (cm) = 10 mm	gram (g) = 1000 mg	litre (l) = 1000 ml
metre (m) = 1000 mm	kilogram (kg) = 1000 g	kilolitre (kl) = 1000 l
kilometre (km) = 1000 m	metric tonne (t) = 1000 kg	

The imperial system

The imperial system is another system of measurement, using units such as the inch and the mile. It has mostly been replaced by the metric system, but some aspects, such as weights in pounds/stones and distances in miles, are still commonly used in the UK. The imperial system is considered old-fashioned, and its units are not often used in maths.

The following table summarises some units of length, weight and volume from the imperial system.

Length	Weight	Volume
inch (in)	ounce (oz)	fluid ounce (fl oz)
foot (ft) (plural, feet) = 12 in	pound (lb) = 16 oz	pint (pt) = 20 fl oz
yard (yd) = 3 ft	stone (st) = 14 lb	gallon (gal) = 8 pt
mile (mi) = 1760 yd	ton (t) = 160 st	

Estimation

Estimating in maths is a way of calculating an answer approximately (getting a 'rough answer'). You shouldn't need to use a calculator or any written methods when estimating, even with large numbers or decimal numbers.

For example, an engineer may have a large project that needs completing. They may have a rough idea of the quantity of material required, as this will give them an approximate idea of the cost. Once the engineer looks to progress with the project, they will then need to find the exact cost of the materials. An estimation is a quick and effective way to see whether or not a project is feasible.

Example

It would be difficult to mentally calculate the amount of mild steel for a project if an engineer required 50 × 3.218 m lengths. However, if we round the more difficult number to 3, the calculation is much easier.

So, 50 × 3 = 150 m lengths required.

This would be an approximate answer.

Costings

Cost of material

The cost of material can be calculated using the following formula:

cost of a material = mass of material × cost per unit mass

(or cost of material = area of material × cost per unit area)

Cost of labour

Labour costs also need to be considered when costing up a project:

cost of labour to make a product = labour time × charge rate

Cost of parts

total cost of parts in a product = £ part 1 + £ part 2 + £ part 3 etc.

Cost to make a product

The three calculations detailed above need to be combined to work out the total cost of a product:

total cost to make a product = cost of materials + cost of labour + cost of parts

Profit

The profit that will be made on a product is calculated as shown:

profit = sales price − total cost

Time

In maths, time can be defined as an ongoing and continuous sequence of events that occur in succession, from the past through to the present and then to the future. In engineering, time is used to quantify, measure or compare the duration of events or the intervals between them. It can even be used to sequence events. Time can be calculated in seconds, minutes, hours, days, weeks, months and years.
- 1 minute is 60 seconds
- 1 hour is 60 minutes
- 1 day is 24 hours

Areas

Engineers need to understand how to work out the area of something, as this will allow them to answer the question: how big is it? Imagine you are a structural engineer and you need to work out how many new houses you can fit onto a newly purchased development site. The first thing you might do is find out how much ground space each house will use by calculating the area of the house's **footprint**.

The area of a shape is a measure of the 2D space that it covers. Area is measured in units squared: square centimetres (cm²), square metres (m²) and square kilometres (km²).

> **Key Term**
>
> **Footprint** The area of land a building takes up.

▲ *This is the area of the house's footprint that you would have to work out to calculate how many houses could fit on a piece of land.*

Area of a rectangle

The following formula can be used to work out the areas of rectangles and squares:

A = L × W

where:

A = area

L = length

W = width.

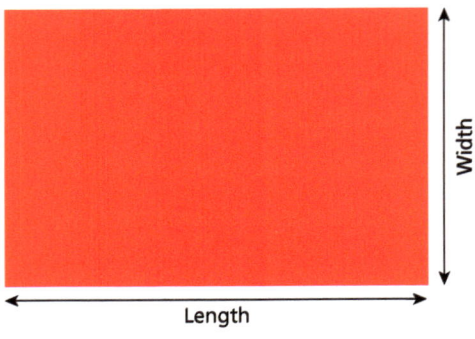
▲ *Calculating the area of a rectangle*

> **Key Term**
>
> **Perpendicular** Something that is at 90° to a given line.

Area of a parallelogram

The area of a parallelogram is found by multiplying the base length (B) by the **perpendicular** height (H):

A = B × H (perpendicular)

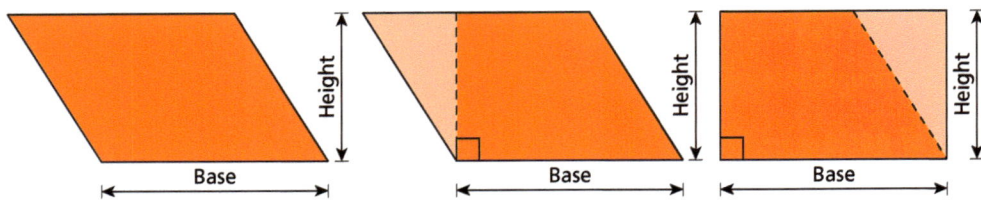
▲ *Calculating the area of a parallelogram*

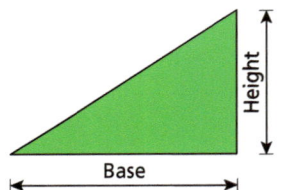 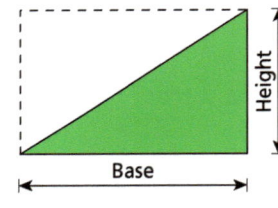
▲ *Calculating the area of a right-angled triangle*

Area of a triangle

If you multiply the base by the perpendicular height, you get the area of a rectangle. The area of a triangle is half the area of a rectangle. So, to find the area of a triangle, multiply the base by the perpendicular height and divide by two, using the formula:

A = (B × H)/2

The principle is also true when calculating the area of a non-right-angled triangle. All you need to do is find the perpendicular line to measure the height. You will then be left with two right-angled triangles. You can work out the area of each and then add them together.

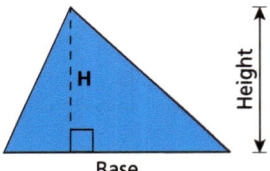 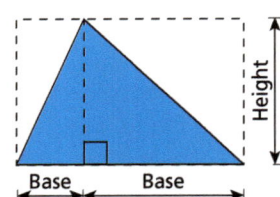
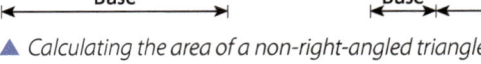
▲ *Calculating the area of a non-right-angled triangle*

Area of a circle

To work out the area of a circle, you need to know the diameter and, from that, the radius. The diameter goes from one edge of the circle, straight through the centre point to the opposite edge. The radius goes from the centre point to the edge of the circle and is half the length of the diameter.

The formula for finding the area of a circle is:

A = πr^2

where:

A = area

π (the Greek letter pi) = a constant with the value 3.14159265359

r = radius.

6 Designing engineering products

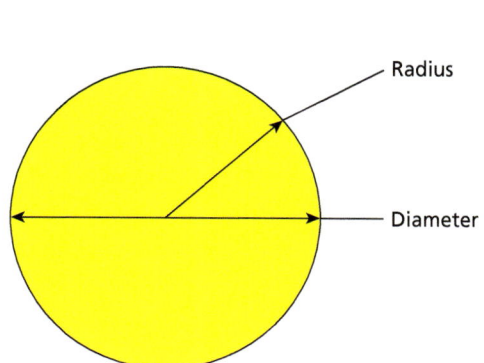

▲ Determining the radius and diameter of a circle

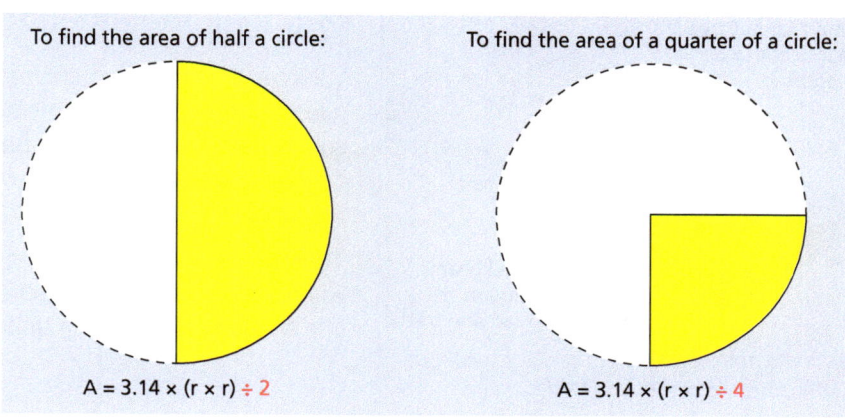

▲ Calculating the area of half a circle and a quarter of a circle

In other words, the area of a circle:

A = 3.14 × (r × r)

(The value for π has an infinite number of decimal places, so we shorten it to 3.14 for use in calculations.)

Area of compound shapes

A compound shape is a non-standard shape. It can look complicated but is actually very simple to work out.

The shape in the figure below looks complicated and you might struggle to work out the area if you are looking at the shape as a whole.

A simple solution to working out the area of compound shapes is to break them up into very simple shapes, work out the area of each shape (using previous examples) and then either add or subtract the individual areas depending on how you broke them up.

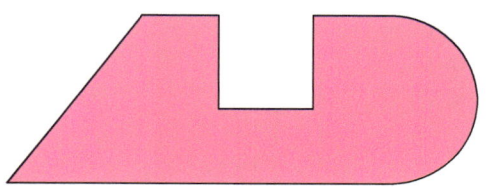

▲ An example of a compound shape

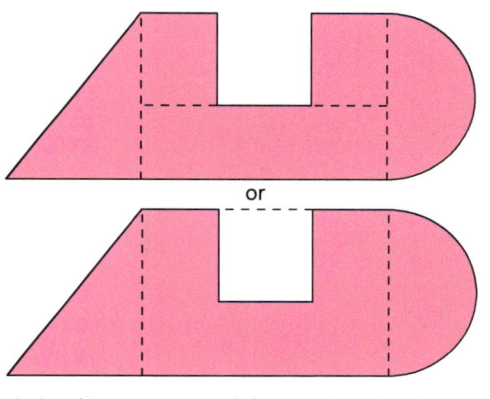

▲ Breaking a compound shape up into simpler shapes makes it easier to work out the area.

Task 6.8

Using your learnt knowledge of areas, work out the area of the following aluminium bracket. If you get stuck, just look back at how to break up compound shapes into simple shapes and then work out the area of each simple shape. Show all your workings out.

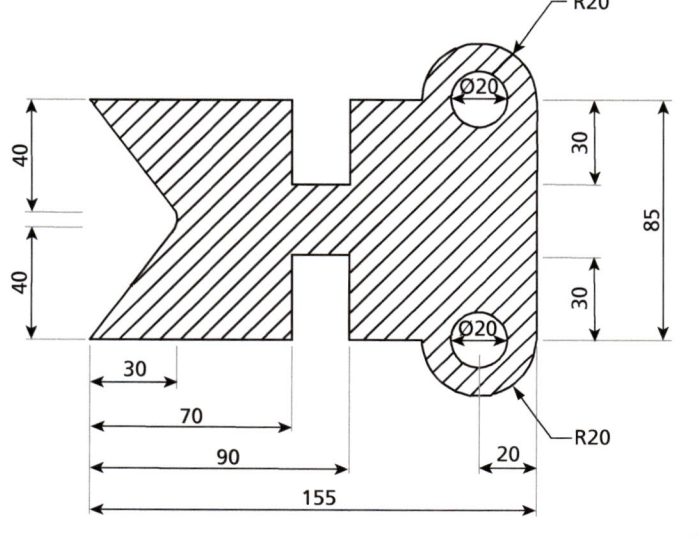

Heavy tarmac (road) that needs to be supported

Steel beams to support the road and traffic

▲ An engineer would need to be able to work out the correct volume of steel for this bridge.

Volumes

Volumes are another mathematical technique that engineers must learn. A good example of this is understanding the volume of steel you might need to support a bridge and all the traffic that will pass over it. Too little steel may cause the bridge to collapse, while too much steel may cause the bridge to be too heavy and too costly to build.

The volume of a shape is the space it occupies in three dimensions, whereas the area only measures the surface (two dimensions). The volume of a shape is measured in units cubed (for example, m^3).

Volume of a cube

The figure on the left is a cubic centimetre. Each of its sides measures 1 cm in length. The volume of a cube is calculated by multiplying the length by the width by the height.

We can write this as:

V = L × W × H

In a cube, the length, width and height are the same. Therefore, the volume of the cube in this example is $1\,cm^3$.

The 3D shape below is made up of a total of eight $1\,cm^3$ cubes. So, its volume is $8\,cm^3$.

Volume of a cuboid

A cuboid is a 3D shape with six rectangular sides. The sides are all perpendicular (at 90°) to adjacent sides. A good example of a cuboid would be a solid brick.

To find the volume of a cuboid, multiply its length by its width by its height. We write this as:

V = L × W × H

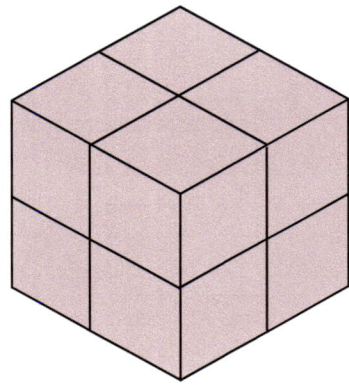

▲ The volume of this cube is $1\,cm^3$.

▲ This shape is made up of 8 identical $1\,cm^3$ cubes, giving it a volume of $8\,cm^3$.

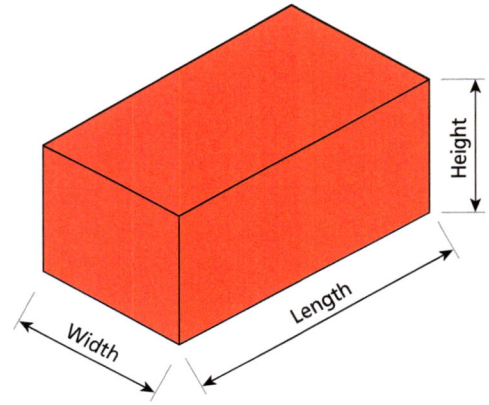

▲ Calculating the volume of a cuboid

Task 6.9

How many cubes are there in the following two shapes? Write down the volume of each.

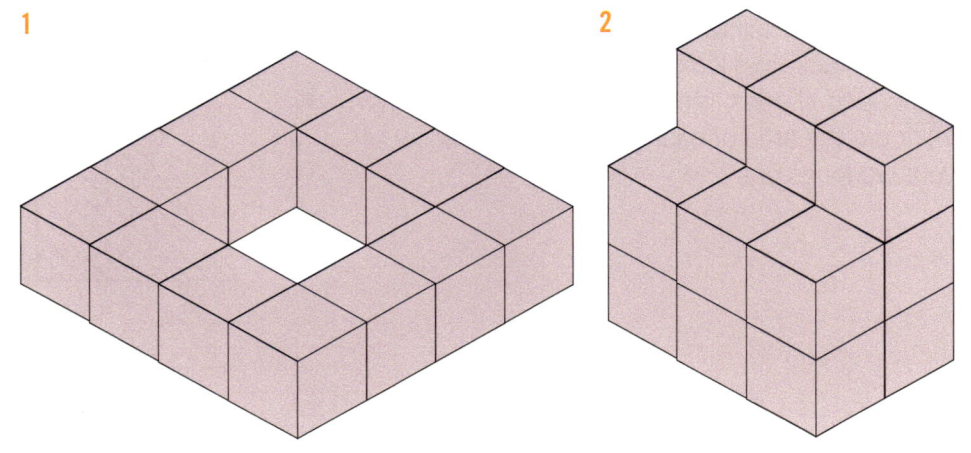

6 Designing engineering products

Volume of a prism

You have already learned how to work out areas. If the area was the end of a section of mild steel (e.g. a square section mild steel as in the figure below) that would also be known as a **cross-section**. The shape of the length of steel could be described as a prism.

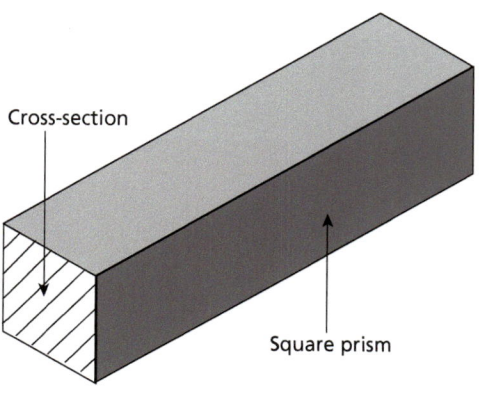

▲ An example of a square prism

This formula works for all prisms:

Volume = Area of cross-section × Length

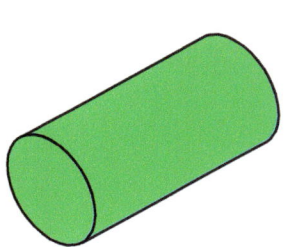
Volume of a cylinder
= Area of circle × Length

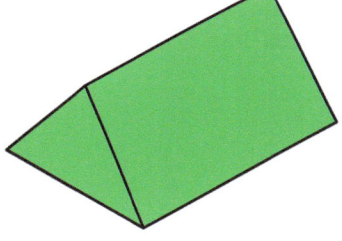
Volume of triangular prism
= Area of triangle × Length

Volume of 'L'-shaped prism
= Area of 'L'-shape × Length

▲ Different types of prism

> ### Task 6.10
> Using your learnt knowledge of areas and volumes, work out the volume of the steel rivet in the figure. Show all your workings out.
>
>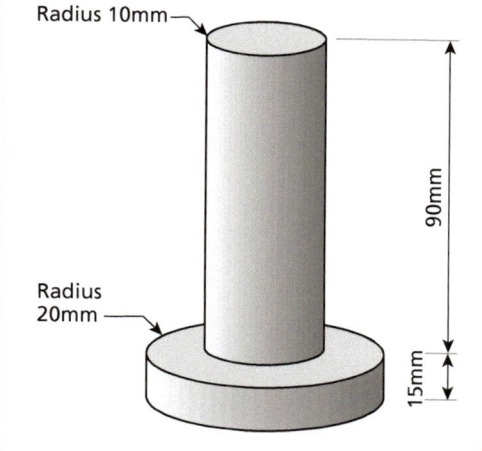

> **Key Term**
>
> **Prism** A solid shape that has two ends of the same shape and size. The length of a prism can vary.
>
> **Datum** A fixed starting point of a scale of operation.

Measuring using datum points

A datum point is a reference used as a starting point to measure from. An engineering datum, used in geometric dimensioning and tolerancing, is a feature on an object used to create a reference system for measurement. In other words, a datum is a reference point, surface or axis on an object against which measurements are made.

We can use the example of a car wheel. The lug nut holes define a bolt circle that is a datum from which the location of the rim can be defined and measured.

Scale

Scaling is a procedure used by engineers to draw an image that is proportional to the actual size of the object. Scaling in geometry means that we are either enlarging or shrinking figures so that they retain their basic shape. Scaling objects is a great way to visualise large, real-world objects in a small space or magnify small objects to make them easier to see.

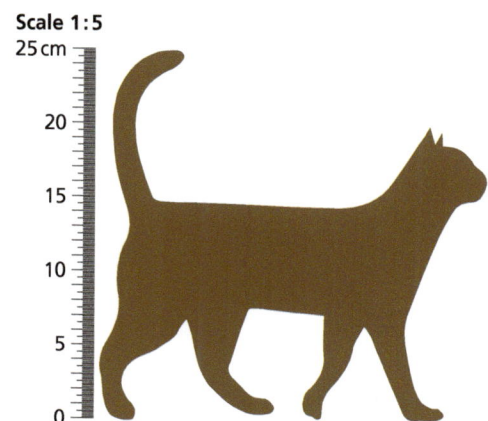

▲ If a cat with a height of 25 cm in the real world is represented as 5 cm in a drawing, it shows that a scale of 1:5 has been used.

A full-scale object will have a scale factor of 1:1. A scale of 1:5 means that the size of 1 unit in the drawing would represent 5 units in the real world.

Scaling up

To scale up means to enlarge a small shape into a large one. The scale factor for upscaling is always greater than 1.

An example of a scale factor of a drawn square with a scale factor 10:1 can be seen below.

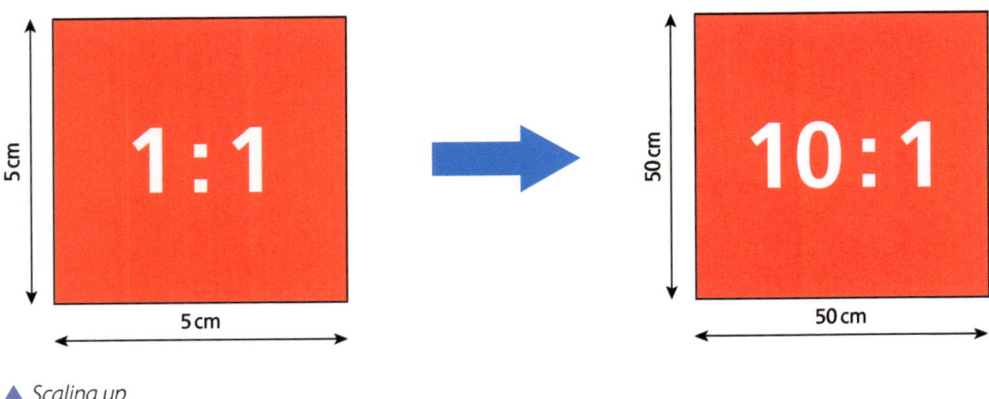

▲ Scaling up

Scaling down

To scale down means that a large number is reduced to a small number. The scale factor for scaling down is always less than 1. A scale of 1:2 shows that the triangle has reduced by half.

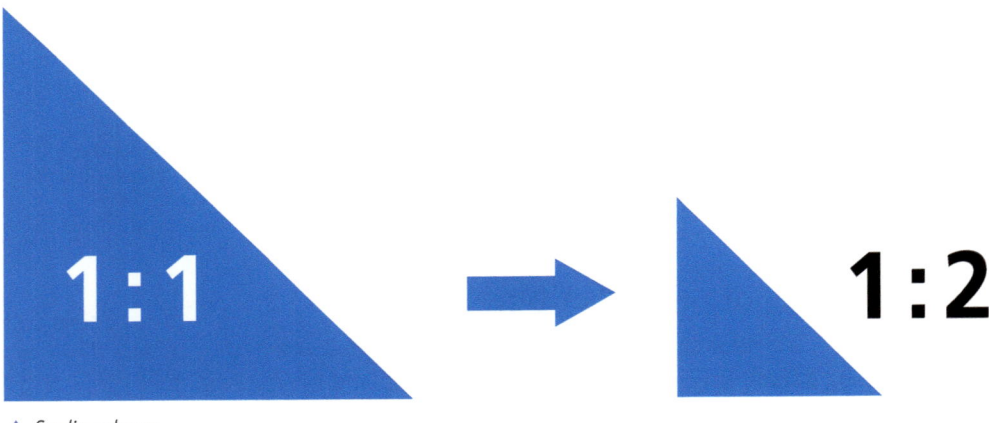

▲ Scaling down

6 Designing engineering products

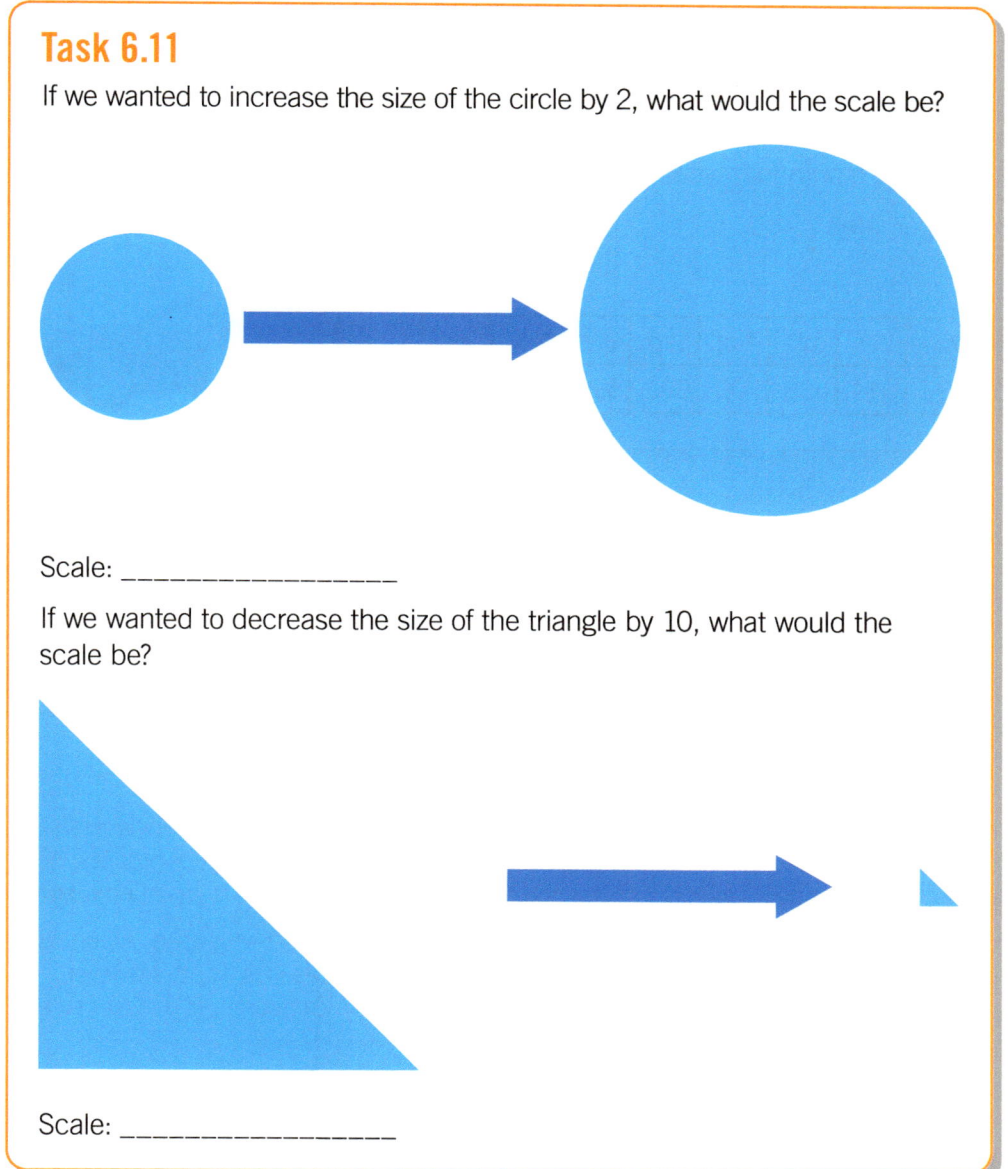

Task 6.11

If we wanted to increase the size of the circle by 2, what would the scale be?

Scale: _____

If we wanted to decrease the size of the triangle by 10, what would the scale be?

Scale: _____

The mean (average)

The mean is the most common measure of average. If you ask someone to find the average, this is the method they are likely to use.

To find out the average from a set of data, add all the data (numbers) together and then divide the result by the total amount of numbers (data).

Below is a set of scores taken from students' engineering tests (11 students, 11 tests, 11 scores):

9, 13, 9, 11, 9, 13, 11, 9, 10, 8, 11

Added together the total is: 113.

So, the mean is: 113 ÷ 11 = 10.27

Now we know that the mean/average score for the test for that particular set is 10.27 per student.

Task 6.12

A family has bought a car. They want to know how much it is costing to run the car in fuel costs per month. They want to know the mean cost and they have recorded the miles travelled each month for a year.

The new car does 10 MPL (miles per litre).

A litre of fuel costs £1.50.

Month	Jan	Feb	Mar	Apr	May	Jun	Jul	Aug	Sep	Oct	Nov	Dec
Miles	600	723	650	760	667	544	556	700	801	599	655	745

Key Terms

Joules (J) The unit used to measure energy or work.

Efficiency The state or quality of being efficient, or able to accomplish something with the least waste of time and effort. In scientific terms, efficiency is the relationship between the total energy input to a system and the useful energy output. The higher the useful energy output, the more efficient the system. Efficiency is calculated as a percentage.

Efficiency

All engineers need to contend with and understand the use of energy. Sometimes it is the use of energy to produce a product, at other times it could be the use of energy by a finished product they have designed.

Engineers deal with different types of energy such as:
- heat
- light
- kinetic (movement)
- chemical
- electrical
- sound
- gravity
- elasticity.

Energy is measured in **joules (J)**. When looking at products and processes, engineers need to understand how much energy would be needed to run a product or process, how much of that energy is used and how much of that energy is lost. This is called **efficiency**.

Look at the figure below. You can see a simple filament lightbulb being used. An amount of electrical energy is used to run the lightbulb. The desired result is 'light'. However, while the lightbulb is on, some energy is being lost as 'heat' and some, to a lesser extent, as 'sound'.

To work out the efficiency of a product/process you can use the following formula:

$$\text{efficiency (\%)} = \frac{\text{(useful energy OUT (J))}}{\text{(total energy IN (J))}} \times 100$$

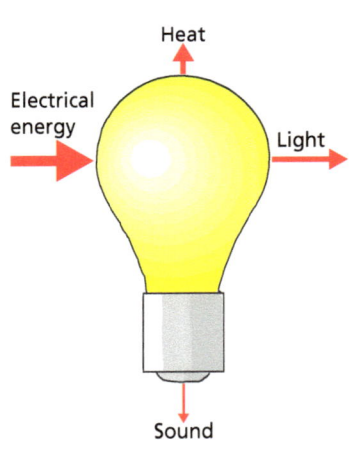

▲ If you add values to each of these forms of energy, you can work out the efficiency of the lightbulb.

Task 6.13

Calculate the efficiency of the lightbulb in the diagram.

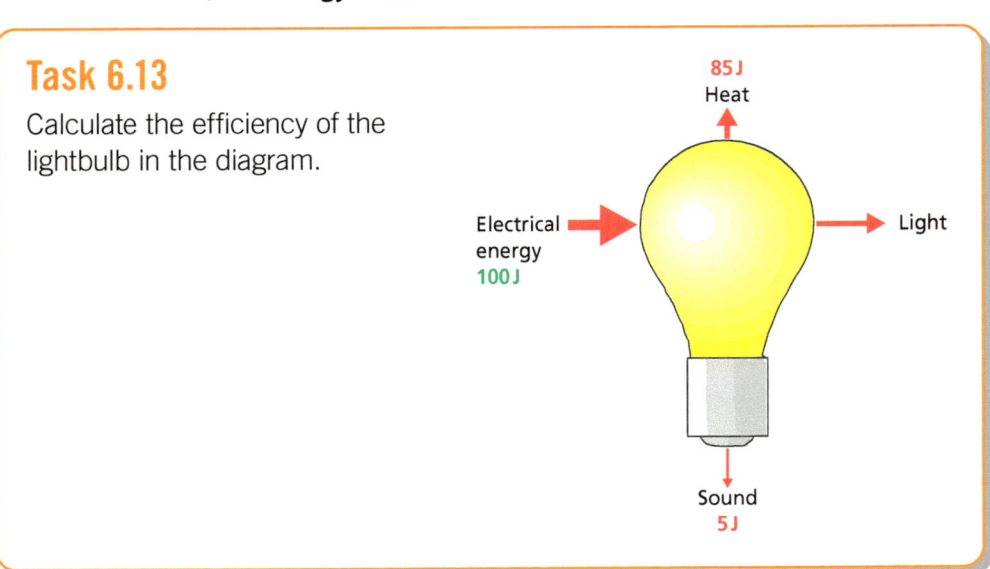

6 Designing engineering products

Mathematical techniques for electronics

Having a basic understanding of simple circuits should be part of any engineer's skill set, as electronics and circuits are now used in many different products.

In this section you will need to understand the basic principles of voltage, current and resistance.

Ohm's law

Ohm's law is a formula that is used to work out the resistance in a circuit. Three elements of a circuit are:

- **voltage** (the power supply of the circuit, the 'push', for example a 9-volt battery)
- **current** (the flow of electricity running around the circuit)
- **resistance** (a measure of the opposition to the current in a circuit; in other words, how the current is slowed down by, for example, the wires and components).

Ohm's law shows how these three elements are connected. In its simplest form, it is written as

V = IR

where:

V = voltage (in volts, V)

I = current (in amps, A)

R = resistance (in ohms, Ω).

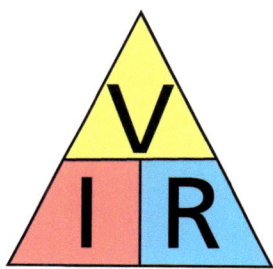

The formula triangle shown will help you to rearrange the formula to work out the missing values.

To work out voltage, you would cover the V with your finger to leave I × R:

voltage = current × resistance

To work out current, you would cover I to leave V/R:

$$\text{current} = \frac{\text{voltage}}{\text{resistance}}$$

To work out resistance, you would cover the R to leave V/I:

$$\text{resistance} = \frac{\text{voltage}}{\text{current}}$$

The figure below is a simple series circuit with four component parts.

Using Ohm's law, calculate the current (I) of the above circuit, using the formula:

$$\text{current (I)} = \frac{\text{voltage}}{\text{resistance}}$$

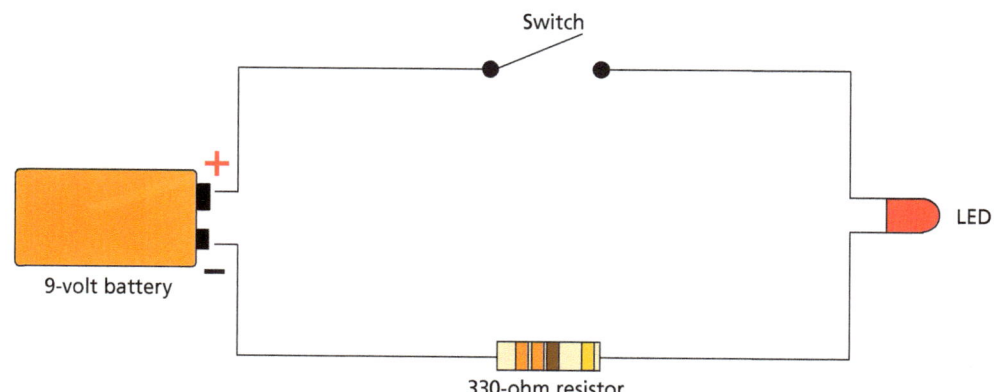

Task 6.14

Using Ohm's law, solve the following problems:
1. A car has an internal cabin lightbulb to light up the interior. The lightbulb has a resistance (R) of 24 Ω. The current (I) running through the lightbulb is 0.5 A. What voltage is needed to run the lightbulb?
2. A small keyring torch is run with a current of 0.3 A taken from a 3-volt battery. What is the resistance of the circuit in the keyring?

Mechanical advantage

Mechanical advantage is a measure of how much a force is increased by using a mechanism, such as a tool or machine, that supports a person to move a weight. It is equal to the force (in newtons) exerted by the tool or machine divided by the applied effort.

mechanical advantage = load ÷ effort

Mechanical advantage using levers

A lever is a simple machine or mechanism that comprises a rigid beam and a fulcrum. The fulcrum is the point at which the beam pivots.

There are three classes of lever: first class, second class and third class. They differ in terms of the position of the fulcrum, load (output force) and effort (input force) in relation to each other.

First-class levers

In a **first-class lever**, the fulcrum is between the load and the effort. If the fulcrum is closer to the load, then less effort is needed to move the load. If the fulcrum is closer to the effort, then more effort is needed to move the load. This idea is illustrated in the figure. A car jack and a crowbar are examples of first-class levers.

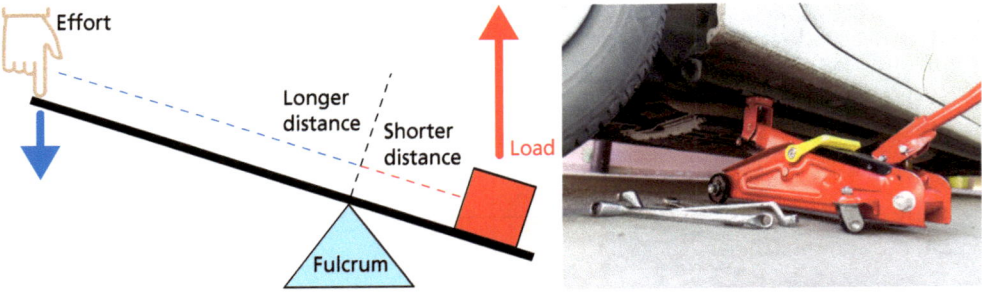

▲ A first-class lever

Second-class levers

In a second-class lever, the load is located between the effort and the fulcrum (see figure). An example of a second-class lever is a wheelbarrow.

6 Designing engineering products

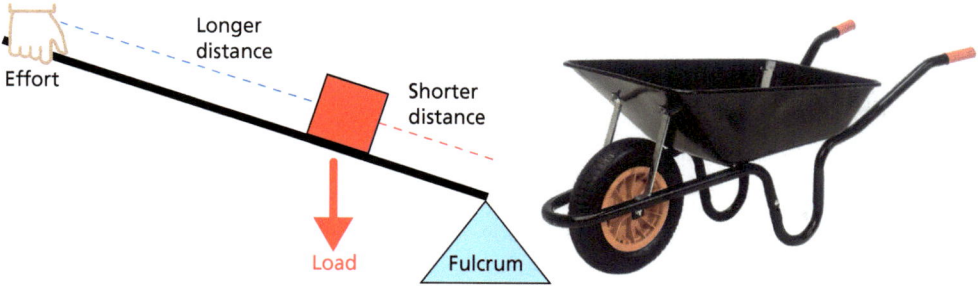

▲ A second-class lever

Third-class levers

In a third-class lever, the effort is located between the load and the fulcrum. Examples of third-class levers include a pair of tweezers or a staple remover.

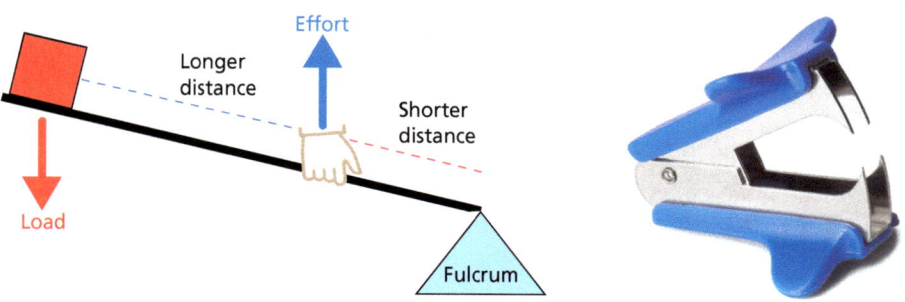

▲ A third-class lever

Task 6.15

Calculate the mechanical advantage of the following levers using the formula:

mechanical advantage = load ÷ effort.

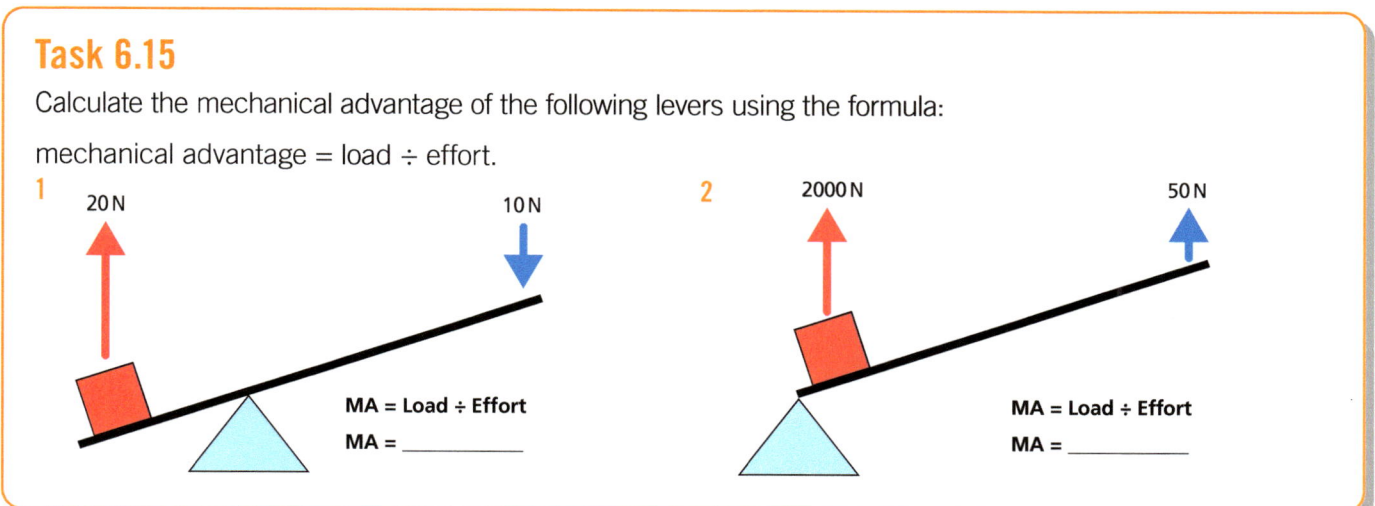

1. MA = Load ÷ Effort
   MA = _____

2. MA = Load ÷ Effort
   MA = _____

Mechanical advantage using pulleys

A **pulley**, at its core, is a simple machine consisting of a wheel with a groove along its circumference. When a flexible rope, cord, cable, chain or belt is threaded through this groove and wrapped around the pulley, it creates a mechanism that can be used to transmit energy and motion.

Pulleys are used singly or in combination to transmit energy and motion. Pulleys with grooved rims are called **sheaves**.

The concept of **mechanical advantage** in pulleys is rooted in the ratio of the load force to the applied force.
- By using pulleys, the force required to lift a load is reduced, resulting in a mechanical advantage that makes the lifting process more efficient.
- The mechanical advantage is influenced by the number and arrangement of pulleys in the system. More pulleys generally lead to a higher mechanical advantage.

In summary, pulleys enhance mechanical advantage by redistributing and reducing the force needed to lift a load, making them essential components in various applications where efficient energy transmission is crucial.

In belt-drive systems, pulleys are affixed to shafts at their axes, and power is transmitted between these shafts by means of endless belts running over the pulleys. The belts provide a flexible and efficient means of transferring power from one pulley to another, facilitating various applications such as machinery operation and vehicle propulsion.

How pulleys facilitate lifting loads
- Pulleys function by distributing the force applied to the rope or belt over a larger distance, allowing for the lifting of heavy loads with less effort.
- The primary advantage lies in redirecting the force needed to lift a load. Instead of applying force directly upward, as one would without a pulley, the force is distributed horizontally along the path of the rope or belt.
- In a basic single fixed pulley system, the load is lifted when a person pulls downward on one end of the rope. The pulley redirects this force upward, effectively making it easier to lift the load.
- When multiple pulleys are used in combination, such as in a block-and-tackle system, the mechanical advantage is increased further. This configuration allows for the distribution of the load's weight among several ropes and pulleys, reducing the force needed to lift the load.

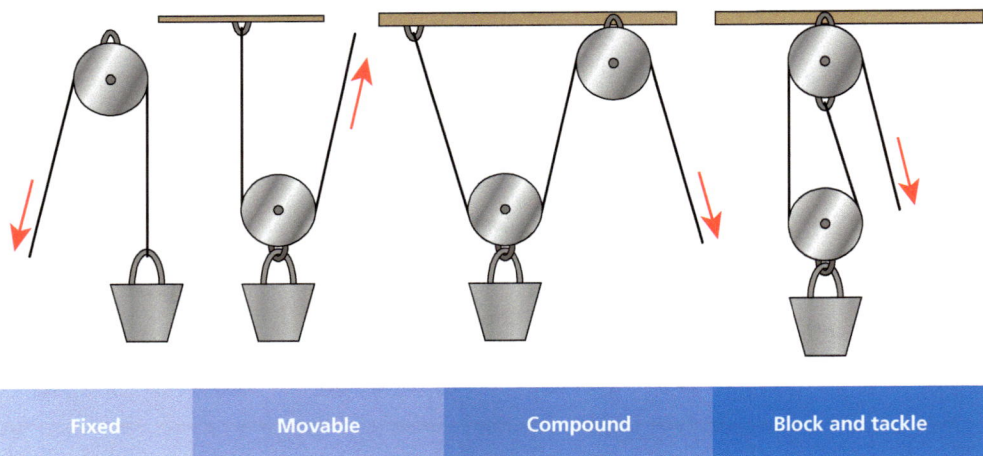

▲ *Different pulley systems*

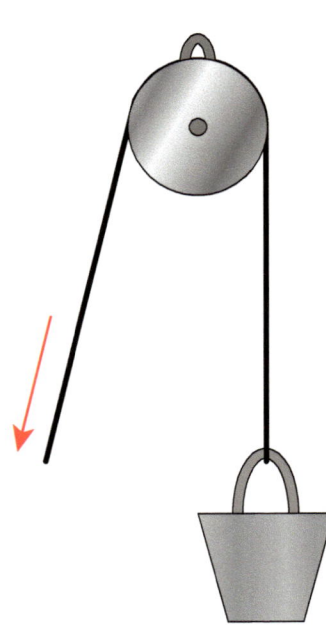

▲ *A fixed pulley*

Fixed pulley
A single-fixed pulley is defined as a pulley with its axis of rotation fixed in place. It is used to shift the direction of the effort.

A single-fixed pulley is characterised by having its axis of rotation fixed in place, serving the primary purpose of redirecting the direction of applied effort. The term 'axis' here refers to the fixed line around which the pulley rotates. However, in the context of a single-fixed pulley, the term 'axis' is not commonly used to describe the rotation point. A more accurate description would be the fixed point or pivot around which the pulley turns.

Movable pulley

In a movable pulley system, the pulley is directly attached to the object being moved. One end of the rope is affixed to a fixed point, while the other end is free. This configuration allows the pulley to move along with the load, and as a result, the applied force is effectively halved, making it easier to lift the object. The movable pulley provides a mechanical advantage by distributing the force needed to lift the load between the two segments of the rope.

Compound pulley

A compound pulley system combines a fixed (not moving) pulley with a movable pulley. The mechanical advantage can be greater than with a system where only fixed pulleys are used.

Block and tackle

A block-and-tackle pulley system is a mixed structure consisting of a cable and at least one pulley wheel. It is used to lift **loads**. With the use of several pulleys, the force required to lift a certain mass can be reduced further.

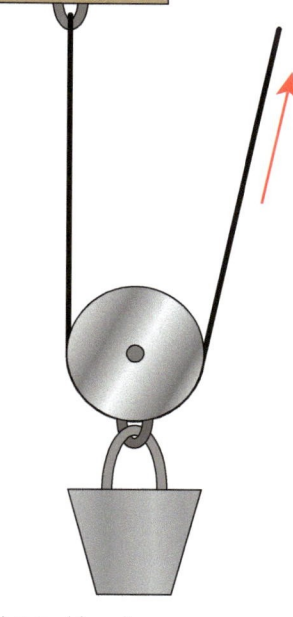

▲ *A movable pulley*

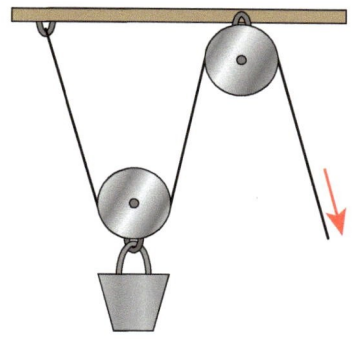

▲ *A compound pulley*

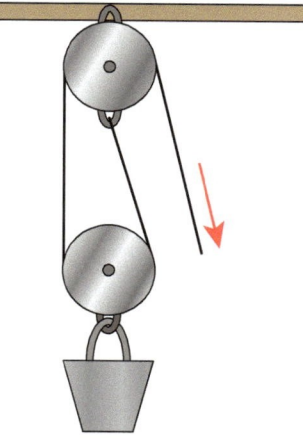

▲ *A block-and-tackle pulley system*

> **Key Term**
>
> **Load** The physical stress on a mechanical system or component.

Example

A pulley is attached to the load and is lifted by pulling up on the rope. However, as there are two ropes effectively lifting the load (sharing the load), the amount of force needed to lift the load is halved. The mechanical advantage is therefore 2. Mechanical advantage can be calculated using the formula:

mechanical advantage = load ÷ effort

100 newtons (load) ÷ 50 newtons (effort)

= 2

The mechanical advantage of this system = 2

For the example pulley system on the right, the load is 200 N. Work out the mechanical advantage and how much effort is needed to lift the load.

Pulley system velocity ratio

The velocity ratio in a pulley system is a measure of the relationship between the velocity (speed) of the effort force applied to the pulley system and the velocity of the load being lifted or moved. It provides insight into how efficiently the pulley system can transmit motion or force from the input (applied force) to the output (load).

> **Key Term**
>
> **Velocity** The speed of something in a given direction.

Introduction to velocity ratio

- The velocity ratio is a crucial concept in understanding the mechanical efficiency of pulley systems.
- It is defined as the ratio of the **velocity** of the effort force to the velocity of the load.
- In simple terms, it helps assess how much the applied force is multiplied or reduced in terms of speed when lifting or moving a load.

Calculating velocity ratio

- The velocity ratio is often expressed as the ratio of the diameters of the pulleys involved in the system.
- For example, in a two-pulley system, if the diameter of the load pulley is twice that of the effort pulley, the velocity ratio would be 2:1.
- Mathematically, the velocity ratio (VR) can be represented as VR = Diameter of Load Pulley / Diameter of Effort Pulley.

Significance of velocity ratio

- A velocity ratio greater than 1 indicates a mechanical advantage, meaning the system multiplies the force applied.
- A velocity ratio less than 1 suggests a mechanical disadvantage, where the applied force is reduced but the speed is increased.
- The velocity ratio plays a crucial role in determining the efficiency and effectiveness of a pulley system for a specific task.

Example

If the velocity ratio is 3:1, it implies that for every three units of distance the effort force moves, the load is lifted by one unit of distance. This indicates a mechanical advantage.

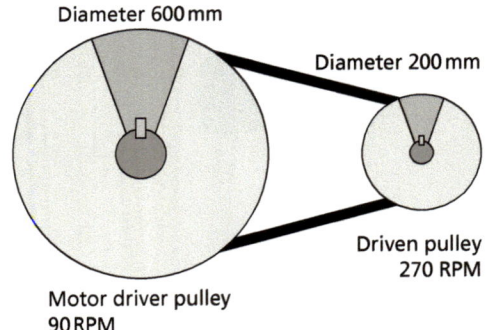

▲ *Pulley system velocity ratio*

Understanding the velocity ratio is essential for engineers and designers when optimising pulley systems for specific applications, ensuring that the system achieves the desired balance between force and speed.

To calculate the pulley system velocity ratio, simply divide the driven pulley diameter by the driving pulley diameter.

Driving shaft
The driving shaft is the component that imparts motion to the connected machinery. It is the shaft that receives power, typically from a motor or another power source, and transmits this power to other components within a system. In the context of a pulley system, the driving shaft is connected to the driving pulley.

Driven shaft
The driven shaft, on the other hand, receives the motion or power from the driving shaft. It is the component that is connected to the load or the part of the machinery that needs to be moved. In the context of a pulley system, the driven shaft is connected to the driven pulley.

In a pulley system, the driving pulley (connected to the driving shaft) and the driven pulley (connected to the driven shaft) work together to transmit motion and force. The velocity ratio, as mentioned earlier, can be calculated by comparing the diameters of the driving and driven pulleys or their rotational speeds.

6 Designing engineering products

Task 6.16

Complete the following velocity ratio calculations for the following pulley systems.

Use the following calculation for the velocity ratio of a pulley system:

$$\text{velocity ratio} = \frac{\text{distance travelled by effort}}{\text{distance travelled by load}}$$

Effort distance = 6 metres

Load distance = 3 metres

Velocity ratio = ?

Effort distance = 12 metres

Load distance = ?

Velocity ratio = 3

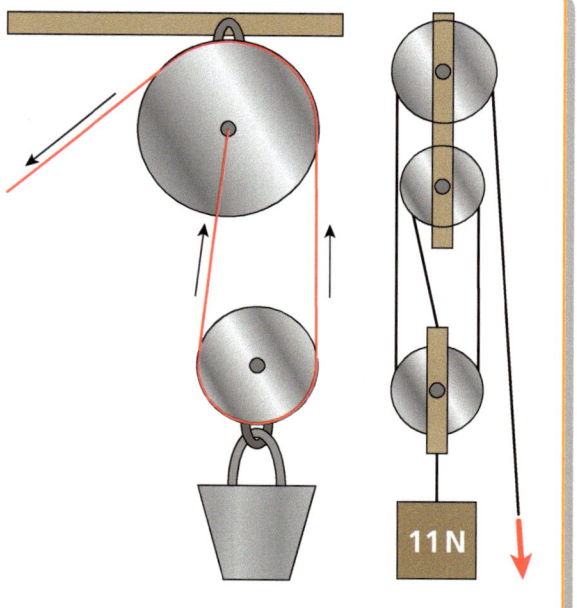

Mechanical advantage and gears

A gear is a mechanical device consisting of a wheel and an axle with teeth around the outside. Gears play a crucial role in transmitting motion and forces within machines. Understanding how gears work is essential for leveraging mechanical advantage in various applications. Gears are often used in combination with other gears to change the direction of forces. The size of the gear determines the speed at which it rotates: the larger the gear, the slower the rotational speed. Gears can be used in machines to increase force or speed.

Basic components of a gear
- **Wheel and axle:** the wheel and axle are the fundamental components of a gear system. The wheel is the circular part, and the axle is the rod or shaft at the centre.
- **Teeth:** gears have teeth around the outer edge of the wheel. These teeth engage with the teeth of other gears, allowing the transfer of motion and force.

Role of gears in changing direction and speed
- **Direction of forces:** gears are often used in combination with other gears to change the direction of forces. When two gears mesh, the rotation of one gear can transfer motion to the other, changing the direction of the force applied.
- **Speed variation:** the size of a gear determines the speed at which it rotates. Larger gears rotate more slowly than smaller gears. This relationship is essential for controlling the speed of various components within a machine.

Driver and driven gears
- **Driver gear:** the gear that initiates the motion is called the driver gear. It is connected to a power source, such as an electric motor or an engine, and sets the entire gear system in motion.
- **Driven gear:** the gear that receives the motion from the driver gear is called the driven gear. The interaction between the driver and driven gears allows for the transmission of motion and force.

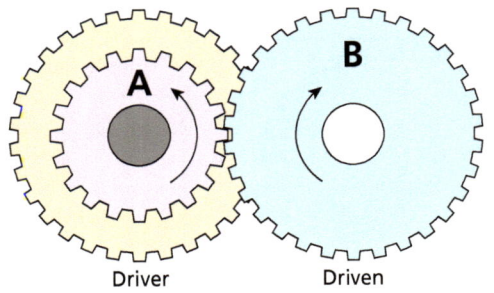

▲ A driver gear provides the source of the power. The driven gear is turned by the driver gear.

Key Terms

Gear train Two or more gears arranged in series.

Parallel Two things are said to be parallel if they are side by side and have the same distance between them continuously. If two lines are parallel, they will never meet.

▲ A spur gear

Mechanical advantage in gear systems:

- Gears can be used to increase force or speed in a machine. When a small gear (pinion) meshes with a larger gear, the larger gear turns more slowly but exerts greater force. This is an example of mechanical advantage.
- Conversely, if a large gear meshes with a smaller gear, the smaller gear turns more quickly but exerts less force. In this case, the mechanical advantage is in favour of speed.

Understanding the principles of gears and their role in transmitting motion allows engineers to design systems that optimise mechanical advantage for specific applications.

An idler gear is used in a **gear train** to control the direction of the driven gears.

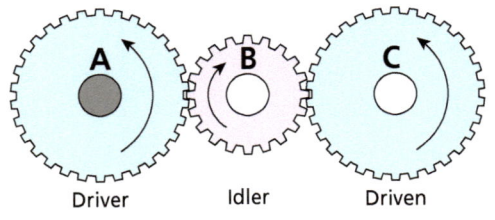

▲ A gear train made up of three gears, including an idler gear.

Types of gears

Spur gear

A spur gear is a cylindrical gear that transmits power through shafts that are **parallel**. The teeth of the spur gear are parallel to the axis of rotation. This causes the gears to produce radial reaction loads (perpendicular to the shaft), but not axial loads (in the same direction as the shaft). Spur gears can be noisier than helical gears (see below) because they operate with a single line of contact between teeth. While the teeth are rolling through mesh, they roll off from contact with one tooth and accelerate to the contact with the next tooth.

Radial reaction loads

Radial reaction loads are forces that act perpendicular to the axis of rotation, typically directed towards or away from the centre of the shaft. In the context of gears, such as spur gears, radial loads arise due to the contact forces between gear teeth. These loads are crucial considerations in gear design and are supported by bearings. Bearings must be selected and designed to handle these radial loads to ensure the smooth operation and longevity of the system.

Understanding and managing radial reaction loads are essential in gear design to prevent issues such as excessive wear, premature failure or deformation of components.

Axial loads

Axial loads are forces that act along the axis of rotation, parallel to the shaft. In the case of gears like spur gears, axial loads are generally minimal. Spur gears primarily generate radial loads, and the design of supporting structures does not heavily consider axial loads. However, in certain gear arrangements or applications, axial loads may need to be addressed to ensure proper system functionality and longevity.

While axial loads are typically not a major concern in spur gears, providing this information adds completeness to the understanding of loading conditions on the shaft.

In summary, radial reaction loads are crucial for supporting the weight and forces exerted by gears on the shaft, while axial loads are generally less significant in spur gear applications but should be considered in specific contexts. The level of detail provided may depend on the audience and the specific requirements of the discussion or application.

Helical gear

Helical gears have teeth that are oriented at an angle to the shaft, unlike spur gears where the teeth are parallel. This causes more than one tooth to be in contact during operation and, as a result, helical gears can carry more load than spur gears. This arrangement also means that helical gears can operate more smoothly and more quietly than spur gears. Helical gears produce a thrust load (a load parallel to the shaft of the gear) during operation which needs to be considered.

▲ *Helical gears*

Double helical gear

Double helical gears are a variation on helical gears in which two helical faces are placed next to each other with a gap separating them. Each face has identical, but opposite, helix angles. Having a double set of helical gears removes thrust loads and allows greater tooth overlap and so smoother operation. Like the helical gear, double helical gears are commonly used in enclosed gear drives.

▲ *A double helical gear*

Herringbone gear

Herringbone gears are very similar to the double helical gear, but they do not have a gap separating the two helical faces. They are generally smaller than the comparable double helical gears and are suitable to be used for high shock and vibration applications. Due to high costs and difficulties in manufacturing, however, herringbone gears are not used very often.

▲ *A herringbone gear*

Bevel gear

Bevel gears are most commonly used to transmit power between perpendicular shafts (those at a 90° angle). They are used where a right-angle gear drive is required. Bevel gears are generally costlier than parallel shaft arrangements and are not able to transmit as much torque for their size.

▲ *A bevel gear*

Worm gear

Worm gears transmit power perpendicularly (at right angles) between shafts that do not intersect. Worm gears produce thrust load and are good for high-shock-load applications but are inefficient when compared to other types of gear. In general, therefore, they are used in lower-horsepower applications.

▲ *A worm gear*

Calculating mechanical advantage

To calculate mechanical advantage is very simple: you divide the output force by the input force and the answer should be a whole number. For a system of gears, the mechanical advantage is given by the gear **ratio**. For gears in a gear train, this is the ratio of the speed of the initial gear to the speed of the final gear. For a simple system of two gears (a driver gear and a driven gear), it can be calculated by dividing the number of teeth on the driven gear by the number of teeth on the driver gear.

$$\frac{40 \text{ teeth (driven gear)}}{20 \text{ teeth (driver gear)}} = \text{velocity ratio of } 2:1 \text{ (gear ratio of } 2:1\text{)}$$

$$\text{Gear ratio} = \frac{\text{Number of teeth on driven gear}}{\text{Number of teeth on driver gear}} = \frac{N_{driven}}{N_{driver}} = \frac{\text{Speed}_{driver}}{\text{Speed}_{driven}}$$

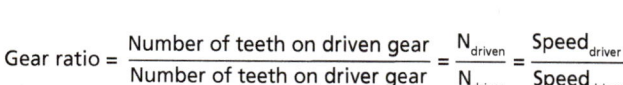

▲ *The formula needed to calculate the gear ratio for a simple system of two gears*

> **Key Term**
>
> **Ratio** The relationship between two groups or amounts that expresses how much bigger one is than the other.

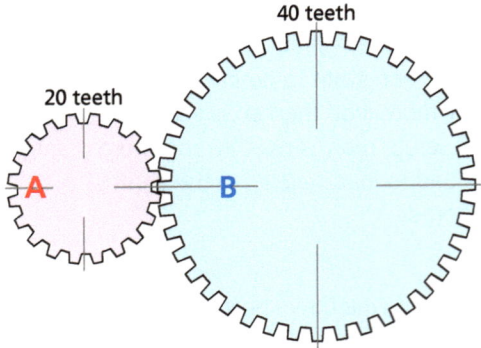

Velocity ratio (VR)

Velocity ratio, often denoted by VR, is a numerical expression representing the ratio of the rotational speeds of the driver gear to the driven gear in a gear system. It is crucial for understanding how the gears transmit motion and influence the speed of rotation.

VR = Rotational speed of driver gear / Rotational speed of driven gear

Why velocity ratio is important

- **Control of rotational speed:** velocity ratio determines how the rotational speed of the driver gear affects the rotational speed of the driven gear. This information is vital for controlling the speed of various components within a machine.
- **Efficiency in power transmission:** calculating the velocity ratio is essential for ensuring efficient power transmission. It aids in designing gear systems that meet specific speed requirements while considering the mechanical advantage and energy efficiency of the system.
- **Optimising mechanical advantage:** velocity ratio is directly related to the mechanical advantage in a gear system. It helps engineers and designers strike a balance between speed and force transmission. Understanding this ratio allows for the optimisation of gear configurations based on the desired system performance.
- **Dynamic control:** velocity ratio plays a crucial role in controlling the dynamic behaviour of a mechanical system. It allows for the fine-tuning of the system's response to changes in rotational speed, ensuring smooth operation and minimising the risk of mechanical issues.

Difference between gear ratio and velocity ratio

While the gear ratio (K_T) represents the ratio of the number of teeth on the driven gear to the driver gear, the velocity ratio (VR) represents the ratio of the rotational speeds of the driver gear to the driven gear.

In an idealised, frictionless system, the velocity ratio is equal to the gear ratio. However, due to factors such as friction and mechanical losses, the actual velocity ratio may differ slightly from the gear ratio.

Example

Continuing from the previous example:

Gear A (driver gear)
- Number of teeth: 20
- Rotational speed: 300 RPM

Gear B (driven gear)
- Number of teeth: 40
- Rotational speed: ?

VR = 300 RPM / Rotational speed of Gear B

Given the gear ratio (K_T) = 2 (40 teeth on Gear B / 20 teeth on Gear A), the velocity ratio can be calculated, and the rotational speed of Gear B can be determined.

6 Designing engineering products

> ## Task 6.17
> Work out the velocity ratio for each of the following gear trains.
>
>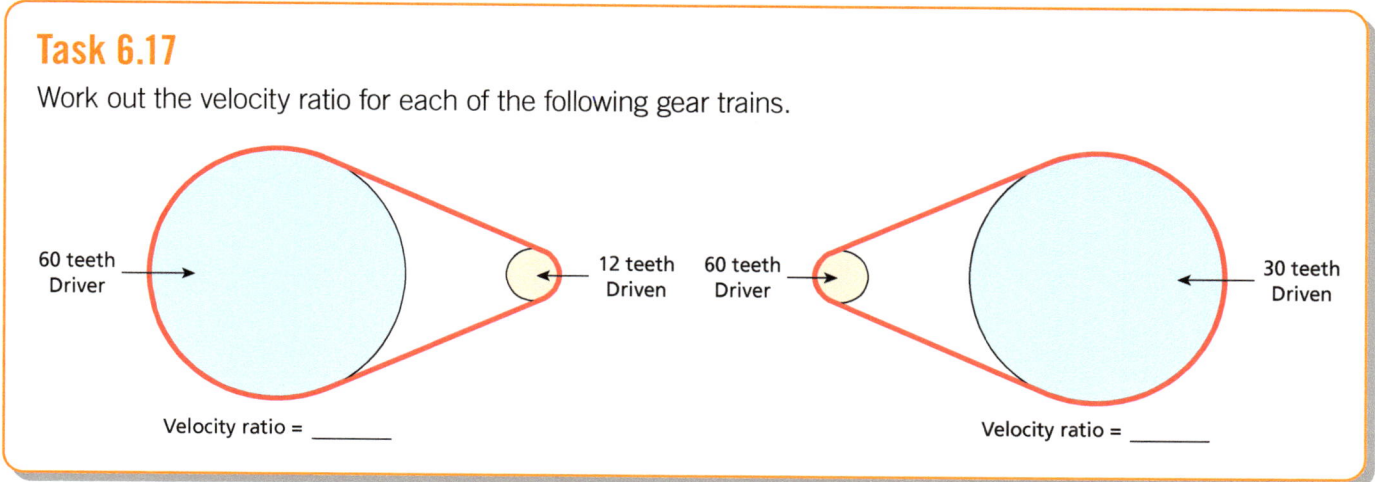
>
> Velocity ratio = _____   Velocity ratio = _____

Displaying data

Graphs

A graph is a type of **chart** used to show the mathematical relationship between varied data sets. The relationship is found by plotting data points on a horizontal (x) axis and a vertical (y) axis. Some common types of graph include bar charts/graphs, histograms and line graphs.

> **Key Term**
>
> **Chart** A way of displaying data or information. It can take the form of a table, a graph or a diagram.

Bar charts/bar graphs

A bar chart or bar graph is a type of graph in which data is displayed in the form of bars with different heights. A bar graph is the graphical representation of categorical data; each bar represents a different category. The height of each bar can tell us how often something happens or show us the number of items in each group. The bars can be drawn horizontally or vertically and must be clearly labelled. There should be spaces between the bars when drawn.

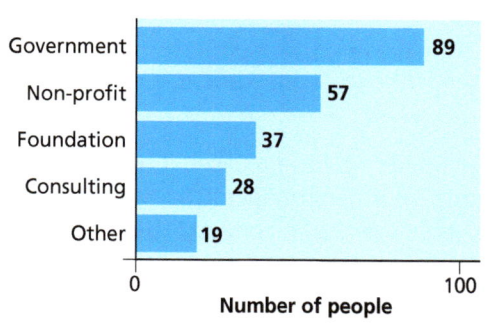

▲ Example of bar chart: horizontal (nominal/categorical)

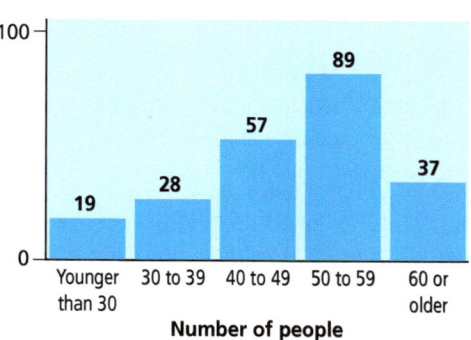

▲ Example of bar chart: vertical (ordinal/sequential)

Histograms

A histogram is the graphical representation of quantitative data. In other words, where there are a range of values for the data rather than discrete categories. There should be no space between adjacent bars.

137

Line graphs

A line graph is used to represent information which changes over time. Such graphs enable trends to be seen clearly. A line graph is generally plotted with points that are joined by a straight line.

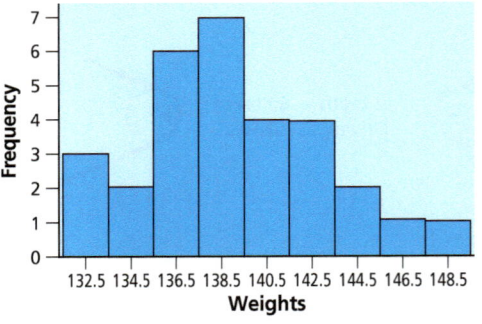

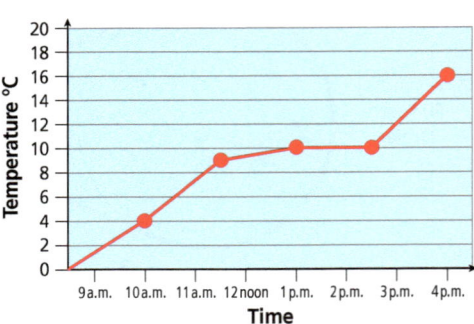

▲ A histogram of weights. In a histogram, the bars must be touching.

▲ Line graph: temperature on 1 September

Pie charts

Pie charts can be used to show percentages of a whole and often represent percentages at a set point in time. Unlike bar graphs and line graphs, pie charts do not show changes over time.

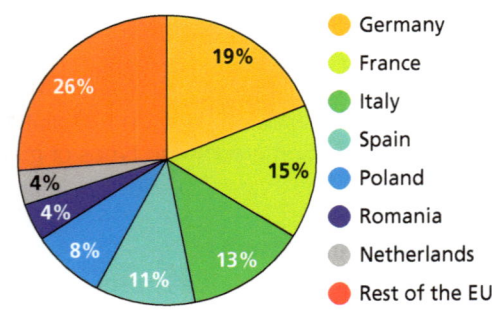

▲ Example pie chart: population of the European Union. Pie charts show percentages of a whole. The percentages must add up to 100.

Evaluation

Introduction

If a human being wants to progress and learn, they have to develop and use evaluation skills.

The first time you crossed a road on your own, you may not have realised it but you almost certainly completed an internal evaluation of your performance. You would have asked yourself: How did I do? Did I judge the speed of the traffic correctly? Did I judge the distances correctly?

By evaluating your performance, you may learn something that allows you to perform better next time. When working on projects, it is good practice for engineers to evaluate their performance, not only for the final solution, but also as the project progresses (correct materials, procedures, finishes, timings, etc.).

When you evaluate performance, you are able to look at your findings and use them to improve the outcome when you are given your next project or brief. You can evaluate on an informal basis by asking yourself questions, asking others their opinions or even visually checking outcomes against your success criteria. However, in this section we will be looking at methods of formally evaluating using existing methods and models.

Evaluating ideas

It can be difficult to evaluate a series of ideas and solutions and evaluate your progress in a project. Where do you start? What do you look for? What should you say?

One of the most common and successful ways is to evaluate against a design brief and specification (the specification being your own list of success criteria). The following are different but accepted methods of evaluating progress and outcomes of projects that an engineer may be required to work on.

Best fit

Look at the ideas/solutions you have created. Compare them to each specification point. Which ideas match the requirements of the specification points best? Which idea meets the most specification points? (Could this be the best idea?) Which idea meets the fewest amount of points? (Should you stop developing this idea?) You could even do this in a table format, as shown below, to keep it formal.

	Idea 1	Idea 2	Idea 3
Specification point 1	✗	✓	✓
Specification point 2	✗	✓	✓
Specification point 3	✓	✗	✓
Specification point 4	✗	✓	✓
Total	1	3	4

Top tip

You can also use set criteria (headings) to help you break down the task into easy-to-manage sections.

By using this method, you can clearly see that the idea that best fits the specification is Idea 3. The table suggests that Idea 3 would be the optimum choice to develop.

Design requirements/fit for purpose

By extracting the key features of a design brief, you are able to quickly identify the design requirements needed in your solution.

When you have identified the design requirements, you can then compare your solution to the key features of the design brief and check to see whether your solution is fit for purpose.

You can also perform this evaluation task by drawing a simple table and checking off the design requirements, as shown below.

Design requirements (design brief key features)	Solution outcome	Fit for purpose?
Must be portable	Yes, as it is lightweight	✓
Must fit into one hand	Yes, as the size is small enough	✓
Must have a range of colours	Yes, as it is made from ABS plastic	✓
Must be easy to charge	No, as it needs a computer to charge	✗

Evaluating with this method helps to identify areas that need addressing and you can go back to your solution and further develop it to meet all the design requirements.

Constraints

Look at your range of solutions and ideas. Which one is going to offer the most constraints? What limitations are you working with?

It might be that your workshop or manufacturing plant only has access to limited manufacturing equipment. Maybe you have a short time to create the solution or even need to learn new skills to be able to complete the project. A big constraint in industry is the overall cost of a project and whether or not some potential ideas are viable to run.

Think about all the constraints and ask yourself these questions:
- Will it be difficult to make?
- Is it too expensive to make?
- Are there enough facilities available to be able to make it?
- Do I have the skills to make it?
- Is there similar competition or is it too close a copy?
- Will the idea have to be compromised/developed too much to make it work?

Top tip

By answering these types of question, you will have a good idea of which ideas would not be viable to continue with and this helps you narrow down the list of potential solutions.

Reliability/operating performance

You can also evaluate the potential reliability and operating performance of your list of solutions. You can try to predict how well they will function during their working life, with the end-users/target market and also in the environment in which they will be used. You could ask:
- Which one is likely to be the most reliable?
- Which will be the most functional?
- Will it continue to perform its function for a long period of time?
- Which one would the end-user find easiest to use?

These questions can be answered by analysing the component parts of the solution such as materials (Will it corrode?), sizes (Is it too big or too small?), shape (Is it comfortable to use/does it fit in its environment?), finish (Does it have an easy grip? Is it safe to use? etc.). These questions will help you decide which is the **optimal** solution for the project you are working on.

Key Term

Optimal The absolute best.
Obsolescence A product is created to become obsolete – old-fashioned or out of date.

Although engineers and designers aim to create new product solutions that are the optimal design, some products are intentionally designed to fail. Imagine a razor blade that never went blunt or a vacuum cleaner that never broke down or lost suction. What would happen to the companies that made those products? If customers didn't need to replace their products, would the companies go out of business? Operating performance and reliability of a solution have to be measured against the needs of the client **and** that of the company, which is why some products are designed to fail. In design terms, this is called planned **obsolescence** and there may be a number of reasons why it is done, including:
- shelf-life (fashion/trends/technology)
- time (reliable for a set amount of time)
- serviceability (parts break and the product needs to be serviced).

Top tip

When evaluating reliability and operating performance of solutions, the needs of all the interested parties (customer, company, etc.) need to be taken into consideration.

Generating a range of engineering solutions

Once engineers have completed the research process through the use of abstracting key data from design briefs and design specifications, they can look at generating ideas that will solve the problem that has been presented to them in the design brief.

Using the design specification as a guide, engineers can start developing a range of ideas that meet the specification points. The design specification is essentially a success criterion for the engineer to follow to ensure that the designed solution is suitable for its purpose.

The solutions the engineers provide should be clear and be communicated in a manner that is easy for the customer or client to understand.

Sketches

Sketches are quick drawings that can be in 2D or 3D. They can be rough outlines of key features of the product or they can go into more depth. Sketches are beneficial as multiple drawings can be completed very quickly compared to other methods, thus saving time. As they are quick drawings, they do not necessarily require rendering. Unfortunately, if the engineer's sketching ability is not to a high standard, sketches can look untidy and can therefore be difficult for the client or customer to understand.

Drawings

Isometric drawings

This would be the preferred method for you to use when completing Unit 1 Manufacturing engineering products, as isometric drawings will hit multiple marking criteria. They are beneficial for this unit as they can provide customers, clients or, in this case, moderators, with more technical information in a format that is neater and easier to read than sketches. Isometric projections allow you to explore more features and parts of your drawing, which in turn makes for detailed annotations and so better communication of ideas. To complete good quality isometric projections, you will require a good understanding of how to produce these designs. Isometric drawings can take more time to produce than sketches. They can be completed by hand, using isometric grid paper or a set square. They can also be produced using CAD software, but this will require a good level of skill. See page 4 of Chapter 1 for more on isometric projections.

> **Top tip**
>
> Sketches should be a quick and rough generation of ideas. Get as many down as quickly as possible. They can always be improved and adapted at a later date.

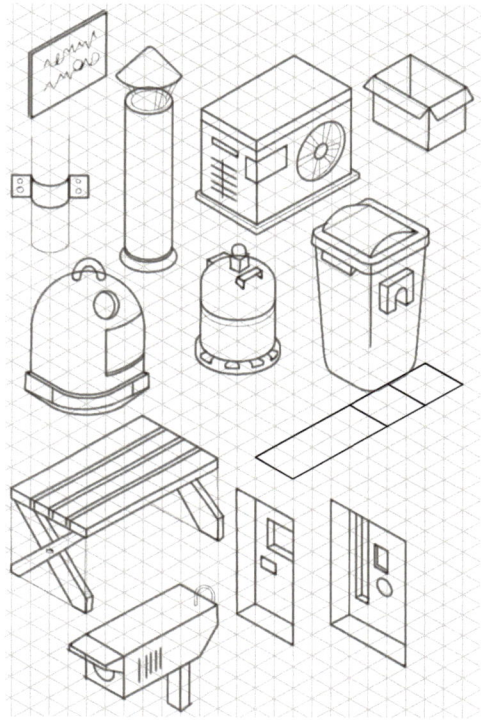

▲ *Design drawings on isometric grid paper*

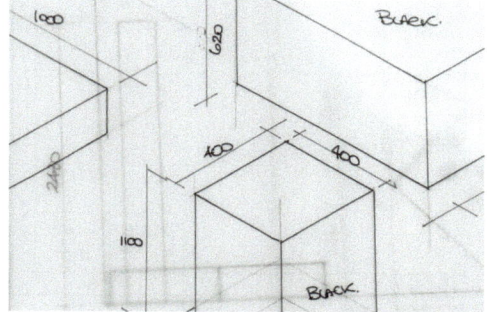

▲ *Isometric drawings produced with a set square*

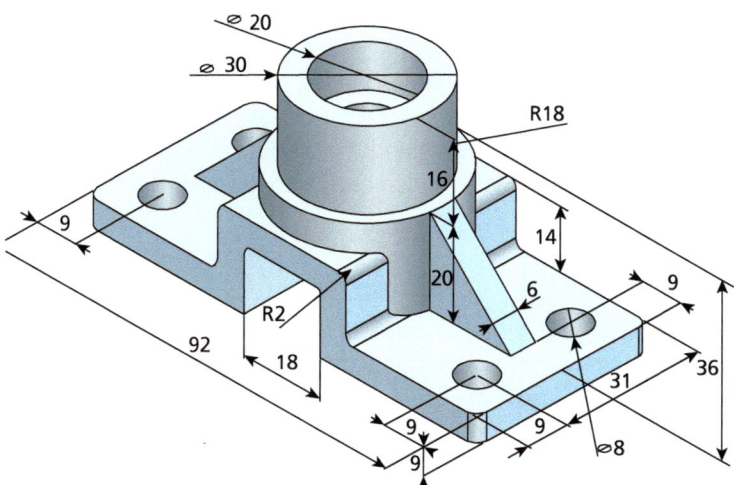

▲ *An isometric projection produced using CAD*

> **Top tip**
>
> Isometric drawings can be produced using a set square or on grid paper. Grid paper can also be used underneath plain paper, so you can trace the 30° lines. Crates (boxes) for each section make the isometric drawings easier to produce.

WJEC Level 1/2 Vocational Award Engineering (Technical Award)

Engineering drawings

Technical/engineering drawings are orthographic projections (see page 13). Engineers would use this method when they have decided on their finished design. This is a technique that you should use to display your final design for Unit 2 Designing engineering products. Orthographic projections allow engineers to communicate all the technical information of their final solution. Another engineer should be able to produce the final design perfectly if provided with the finished orthographic projection.

Orthographic projection is a standard used around the world so that everyone can understand the technical information. Engineers need a high level of skill to create and translate these drawings. The drawings must be accurate enough to ensure that the product is manufactured correctly. Orthographic projections can be completed by hand or by using a CAD program.

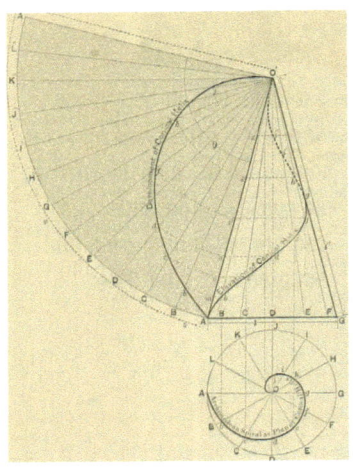

▲ *Third-angle orthographic projection drawn by hand*

 **Top tip**

The engineering drawing will show front, top and side views, along with any measurements required to produce the item.

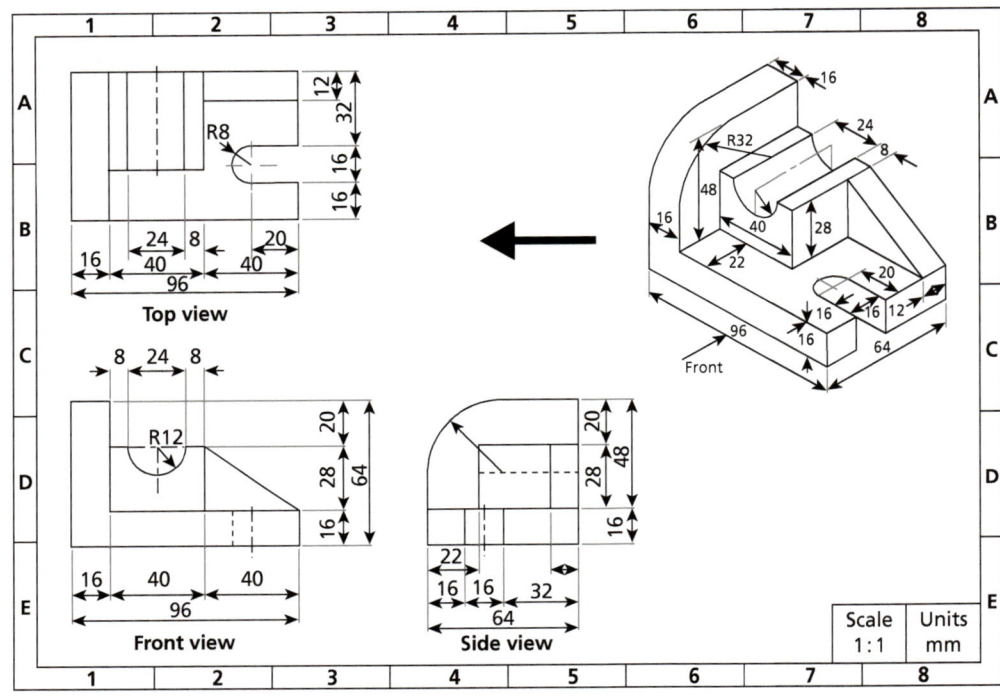

▲ *Third-angle orthographic projection drawn on CAD*

Communicating design ideas

For a successful project outcome, an engineer must be able to communicate their designs in an effective manner. If the design ideas are communicated effectively, it will allow customers, clients or other engineers to understand the product and how it is manufactured. It will also provide further key information on materials, cost and functions.

Communicating designs can be shown in the following ways:

Annotations

Detailed annotations around sketches or isometric projections allow the clients or customers to understand the design more fully. The annotations support the sketches by further explaining the key elements of the design. Engineers could use acronyms such as ACCESS FM (see page 110) to ensure that all key criteria are included in the annotations.

 **Top tip**

Annotations can start with ACCESS FM, but further headings can be added and discussed to communicate the design more effectively.

6 Designing engineering products

Dimensions

Dimensions are measurements and they are an essential part of an engineer's technical drawings or sketches. They provide a quick, easy and effective way of showing the size of the designed product. It should be easy for the client or customer to interpret this information.

Top tip

As shown in the figure, when dimensioning, arrow heads need to be blocked out, lead lines must not touch the object. The convention when writing dimensions is that you should stick to one side where possible. For example, if you start adding your dimensions to the bottom, you must attempt stay at the bottom.

▲ Dimensions are essential in a technical drawing.

Detailed sketches of key parts

Another way of communicating ideas clearly is to create detailed sketches or drawings of key components. This is a quick and effective way of showing how your design functions and is manufactured. This type of visual representation may help customers or clients to understand the product better. Some people may find detailed sketches easier to follow and understand than annotation.

Top tip

Use isometric crates and guidelines to allow you to produce detailed sketches of parts, or highlight a section of the drawing and draw it separately.

Developing ideas through to a conclusion

Once an engineer has generated a range of initial designs, they should then look to develop them further, using their specification points as success criteria. An engineer will select what they consider their best design, with regards to meeting the client or customer requirements.

Developing ideas could be as simple as changing shapes, colours or adding additional features and functions. A developed idea should show progress from the initial design stage.

SCAMPER

The SCAMPER technique is a brainstorming technique used to develop or improve products or services. SCAMPER is an acronym for Substitute, Combine, Adapt, Modify/Magnify, Put to other use/Purpose, Eliminate and Reverse/Rearrange.

You can use this technique to improve the quality of your design idea. It is a handy acronym which will provide key headings for you to consider and discuss.

The table below gives an example of how a phone holder/stand has been developed through the use of SCAMPER.

SCAMPER	Questions	Example: a phone stand
Substitute	What material or pieces can you change? Can you change the purpose of the design?	Could it be made from acrylic, metal, timber or other materials?
Combine	Can you combine this product with another to make something new?	Could you combine the stand with a charger and create a wireless phone charger?
Adapt	Which parts of the product could be adapted to change the nature of the product?	Could you create space to manually charge the phone and make it more useful to the end user?

▲ The phone holder/stand

Top tip

In a similar way to ACCESS FM, the headings in SCAMPER are an extremely useful starting point, but further headings can be added to improve the depth of design development.

SCAMPER	Questions	Example: a phone stand
Modify/Magnify	What can you add to modify the product? Can you change the shape?	Could you increase the size so the stand can hold a tablet?
Put to other use/Purpose	Could the product be used for something else other than the original intention?	Could you add a bracket to hold the phone and an arm to lift the holder off the desk?
Eliminate	How could you simplify the product? What features or parts could you eliminate?	If you took the product to its skeleton and made it from wire could it serve the same purpose?
Reverse/Rearrange	How could you rearrange the product? Which roles could you reverse/swap?	Could you change the layout of the stand by putting the support at the base and brackets at the side to hold the phone in place?

Top tip

Show the journey of your design. Start with your initial design, then show your first improvement. On a further drawing, include your second improvement etc.

Developed drawings

Engineers could develop their ideas through stages. The first drawing would be the initial design from the design generation stage. The second drawing highlights the first improvement added to the initial drawing. The final drawing is of the final design, showing two improvements that have been added to the initial design. This is an excellent way in which an engineer can communicate their design process. It also allows the customer and clients to see how the design has been developed through stages.

CAD models

Engineers can develop and model their designs using CAD. These models can be generated quickly and can be computer tested (by means of **FEA**) to ensure they function correctly before the final design is made. FEA, or finite element analysis, uses computer software to analyse how a design will perform in real-world conditions. The benefit of a CAD model is that it can be edited and adapted; the development process is ongoing. CAD models allow the clients and customers to view all angles of the model, whereas drawings only show the angles that the engineer has produced. The downside to CAD modelling is that engineers need a good level of skill and the CAD software can be expensive to purchase.

▲ An example of a CAD model

Physical models

Engineers can use physical models to communicate and develop their designs. They tend to be produced only when engineers are pretty certain what their final design will be. Physical models allow an engineer to test solutions on their models and determine whether the size or scale is suitable for their problem. Physical models can be made from card, paper, plastics, woods, foam and metal. They can also be produced with CAD/CAM, for example by 3D printing or a CNC box router.

Prototypes

Prototypes are similar to models but they tend to be more accurate and as close as possible to the final solution. Prototypes are often functioning versions of the design and so can be tested in the environment for which the product has been designed. This testing allows the engineer to understand whether the needs of the client or user have been met. The making of a working prototype takes a high level of skill and can be expensive and time-consuming to create. In addition, the correct equipment must be available. Before producing a prototype, an engineer must be fully satisfied that their product hits the design brief and specification points.

Evaluation techniques

At this last stage, when the final design has been completed, engineers should look to evaluate their product outcome. This is an important step in the design process as engineers can identify aspects of the design that went well or aspects that could be further improved. Engineers may look to make amendments to further designs or even look at the manufacturing process. The evaluation of the product will make future products better.

Engineers evaluate using existing evaluation techniques. These techniques are recognised models and are also a useful tool in the engineering/design industry to evaluate projects, outcomes and processes.

As in Unit 1 Manufacturing engineering products (Chapter 4), your design solutions could be evaluated with methods and models such as using a SWOT analysis or measuring against the brief/specification.

> **Top tip**
>
> Physical models do not need to be perfect or accurate in the same way that a prototype needs to be. They can be made from a variety of different materials.

> **Top tip**
>
> SWOT will assess the fundamentals of an evaluation. It does not stop you from adding further headings for depth.

7 Understanding the effects of engineering achievements

In this chapter you are going to:
- understand some different paths of engineering
- recognise achievements of engineers
- understand the positive effects of engineering on day-to-day life and society
- understand how engineering affects the environment
- see how engineers can consider the environment in their work
- look at new engineering developments that have a positive effect on the environment.

This chapter will cover the following areas of the WJEC specification:

Unit 3 Solving engineering problems: 3.1 Understanding the effects of engineering achievements
- 3.1.1 Describing engineering developments   - 3.1.2 Explaining the effects of engineering achievements   - 3.1.3 Explaining how environmental issues affect engineering applications

Describing engineering developments

Unit 3 is the examination and this chapter will cover key information that could be on the examination. However, you should understand that content learned and developed through this book – from Unit 1 Manufacturing engineering products and Unit 2 Designing engineering products – also makes up elements of the examination and should be revised.

Introduction

In this chapter you will be looking at some of the achievements of engineering and seeing how those success stories have had a huge impact on humanity and the world. It is quite easy to dismiss engineering as a career or job where you have to wear overalls, fix something and end up with oily hands. This stereotype of engineering is extremely inaccurate, as engineers are the people who, throughout history, have changed the world with their innovations. For example, an engineer is not the person who fixes an engine, but rather the person who designs and creates the engine. The jobs that require maintenance, service, repair or installation tend to be populated by 'technicians', whereas the careers that involve designing, creating and innovating tend to be populated by 'engineers'.

7 Understanding the effects of engineering achievements

Engineers need the essential abilities and knowledge to create new equipment, machines and structures. They must:
- understand materials and their properties
- know how to use machines, tools and processes
- be able to use new technologies such as CAD/CAM and 3D printing
- understand modelling and prototyping
- have the fundamental skills needed to build their new creations from scratch.

Modern engineers are individuals with huge skill sets and are among the best problem solvers in the world today.

Three areas of engineering have had a significant impact on the world throughout history:
- structural engineering
- mechanical engineering
- electronic engineering.

We will look at these in turn.

▲ *Engineers designing a jet engine.*

Structural engineering

Structural engineers apply their knowledge and skills to create functioning structures such as bridges and buildings, and work as part of a team on different civic projects such as large dams, reservoirs and the Channel Tunnel. Structural engineers can work in partnership with other professionals, such as architects, to create large structures. While an architect might create the overall 'look' of the structure, a structural engineer would specify the materials it would be made from as well as change, amend and develop the design to ensure the structure would function and meet not only the requirements of the client but also conform to all safety standards.

The following sections detail some examples of typical structures that structural engineers work on.

Bridges

A bridge is an engineered structure that spans an obstacle such as a gorge, path or river without obstructing what is underneath the bridge. There are many different types of bridges that exist to perform lots of different functions such as carrying traffic (pedestrians, cars, trains), carrying water (**aqueducts**, canals) or even allowing for manoeuvrability (movable bridges for the armed forces). Engineers have been designing bridges for thousands of years to allow us to live more efficiently in our landscape.

The Romans were among the most accomplished bridge builders in history. They discovered the strength of arches and used these shapes to create some of the most iconic ancient bridges and aqueducts in the world. Many of their engineered structures survive today.

When designing bridges, structural engineers have to understand the forces of **compression** and **tension**. The figure below explains how these forces interact using the example of a simple house roof.

Key Terms

Aqueduct A bridge that carries water from one place to another.

Compression A force that squashes or squeezes things together.

Tension A force that occurs when something is being pulled or stretched.

▲ *The Pont du Gard, in southern France, is an example of a well-built Roman aqueduct that still survives today.*

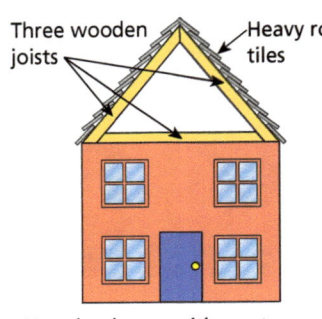
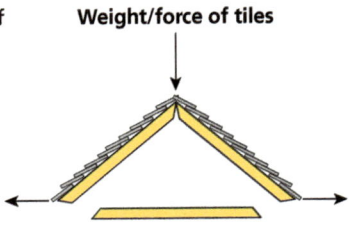
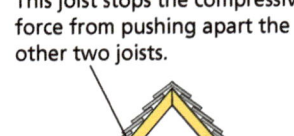
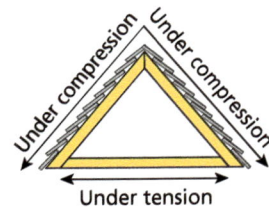

Here is a house with a cutaway of the roof. Notice how the roof has three wooden joists in a triangular shape.

The weight of the tiles puts the roof under **compression** and tries to force apart the joists making the roof collapse.

The third joist stops the two joists that are under compression from being forced apart. The third one is now being pulled apart and is under **tension**.

Here you can see how using simple engineering you can create a roof for a house that will support a lot of heavy tiles.

▲ Understanding the forces of compression and tension

▲ The Millau Viaduct in southern France

The Millau Viaduct

The Millau Viaduct is one of the greatest engineering achievements of modern times. It is a 343-m tall cable-stayed bridge spanning the valley of the River Tarn near Millau in southern France. At the time of writing, it holds the record for being the tallest bridge in the world. It was designed by a team of structural engineers headed up by the British architect Lord Norman Foster.

In a cable-stayed bridge, the central pylon supports the weight of the beam with a series of attached cables. The pylon is generally made from reinforced concrete, as this material performs very well under compressive forces. The cables are made from some form of steel, as steel can withstand the tension forces. Cable-stayed bridges often have more than one pylon to support longer spans.

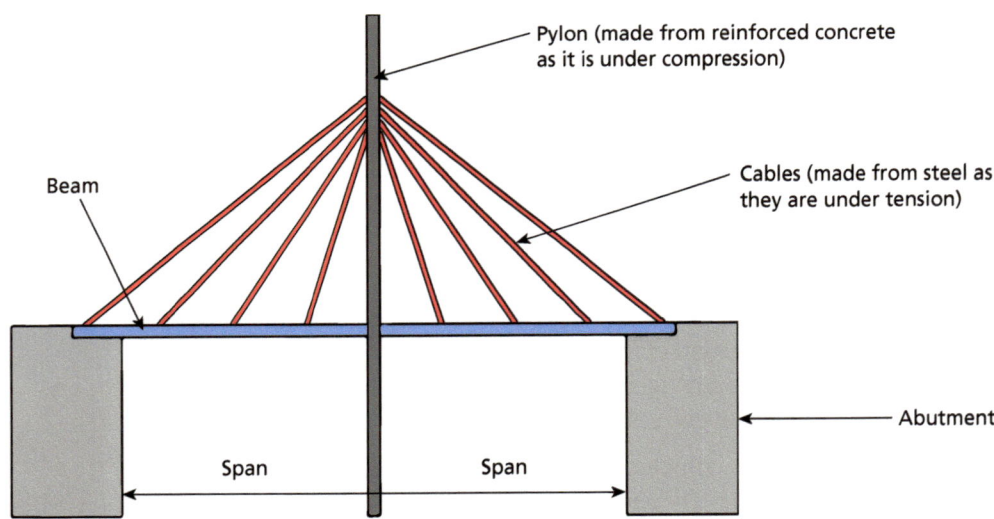

▲ The mechanism behind a cable-stayed bridge

Skyscrapers

Skyscrapers are very tall buildings with many floors. They are called 'skyscrapers' as they are so tall they seem to scrape the sky. They are generally built (fabricated) with a steel frame that supports all the floors and walls. The steel framework bears the load of the rest of the building and all the people, furniture and equipment inside. Skyscrapers are designed and built to withstand extremely high winds, lightning and even earthquakes.

▲ The London skyline, showing some of the UK's tallest skyscrapers

7 Understanding the effects of engineering achievements

When there is no more room to build outwards in a densely populated area such as a city, then you have to build upwards. As technology, engineering and building techniques improve, the number of floors in new skyscrapers increases. At the time of writing, the Burj Khalifa skyscraper in Dubai is the world's tallest building. It has 163 floors and stands at just over 828 metres. This record is set to be overtaken by the Jeddah Tower in Saudi Arabia which, when completed, will stand at around 1 km in height. Skyscrapers allow more space for humans to work and live, but have a relatively small footprint.

How skyscrapers are constructed

Engineers use their knowledge and understanding of materials, properties of materials and forces to construct very tall, very safe buildings such as The Shard, which stands at 310 metres.

The following figure gives a simplified explanation of how a skyscraper is constructed.

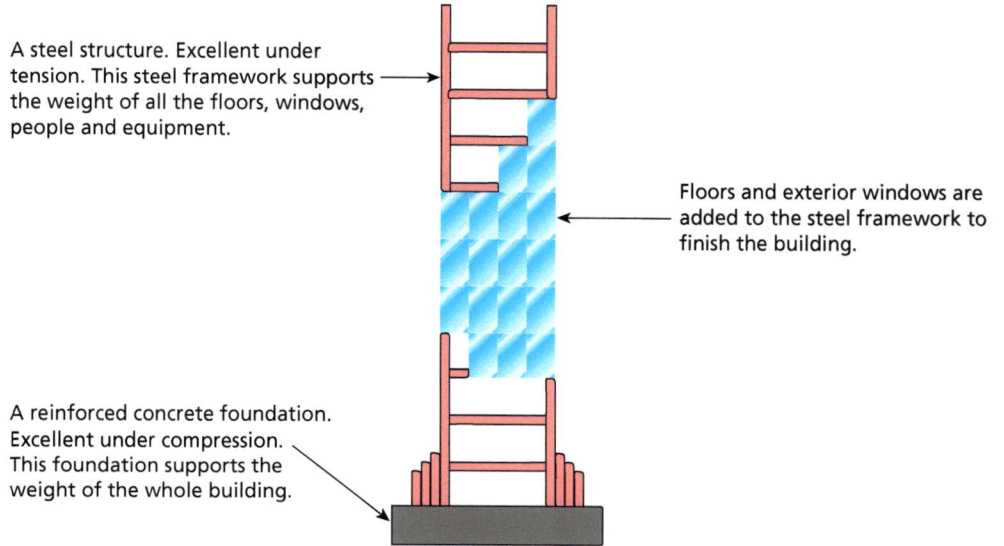

Key Term

Top-down construction
When constructing a building with basement floors you can complete the structure of the higher floors before excavating and constructing the lower floors.

The Shard

The Shard is one of the newest skyscrapers in the UK, having been completed in 2012. It is located in Southwark, London, and, at 310 metres tall and with 72 habitable floors, is not only the tallest building in the UK but the tallest building in Western Europe. It was designed by a famous Italian architect called Renzo Piano and built by a team of structural engineers from Williams Sale Partnership (WSP). Construction of The Shard involved innovative engineering techniques developed by the structural engineering team, such as **top-down construction** and the use of concrete in the middle of the building as well as steel. The Shard contains office spaces, restaurants, a hotel and even apartments.

▲ The Shard is the tallest building in the London skyline.

Task 7.1

Answer the following questions:
1. What modern materials could you use for bridge building?
2. What properties do you need in a material used for building a bridge? Why do the materials need these properties?
3. Name a famous modern bridge.
4. Who designed the bridge you named?
5. What is the tallest skyscraper in the UK?
6. Who were the structural engineers that built it?
7. What materials could be used to build a skyscraper and why?

Mechanical engineering

Mechanical engineers use their knowledge of materials, material properties, mathematics and physics to design and maintain mechanical systems. A mechanical system is a system that uses power (from a source) to perform a specific task.

An example of an early mechanical system is a water wheel. A flour mill, like that illustrated in the diagram, would have used the force of running water from a river as a power source to turn a large water wheel. The water wheel then turns a shaft that is linked to a gear that is linked to a pair of millstones. The millstones grind grain into flour that is used to bake, for example, bread.

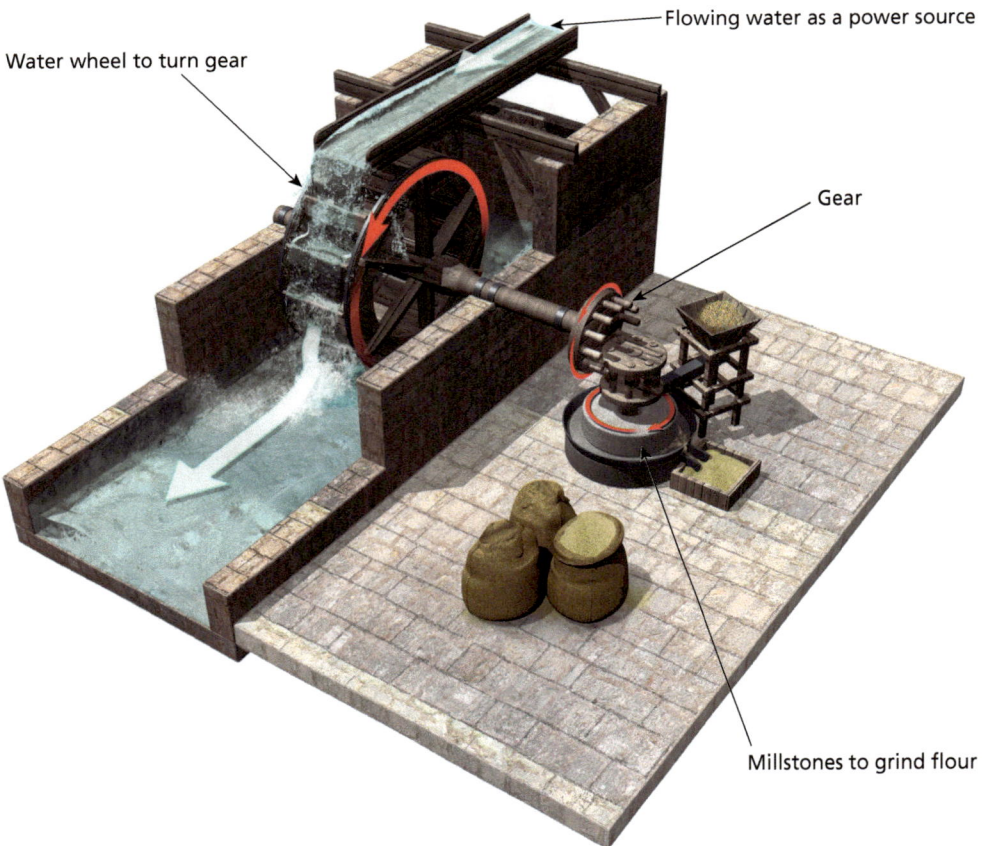

▲ A flour mill using water as a power source

There are many types of mechanical systems that make up different machines which have vastly improved the lives of people and industries. From simple water mills to steam-powered engines to the latest jet propulsion and kinetic energy recovery mechanisms, mechanical engineers are always looking to innovate new mechanical systems for the improvement of society.

Automobiles

A couple of hundred years ago, most people only travelled short distances in their entire lifetime. Long-distance travel was not an option for people reliant on horses and carts or walking. Daily life for the majority of people was restricted to living and working in their local vicinity. However, in today's world of advanced mechanical engineering, the average person can easily travel an average of 12 000 miles a year, because most people have and use automobiles.

The internal combustion engine

The internal combustion engine is the power unit needed to drive most modern cars (with the exception of electric cars). Étienne Lenoir was the Belgian–French engineer who, in 1858, invented the first commercially successful internal combustion engine. What we know as the 'modern' internal combustion engine was created in 1876 by Nicolaus Otto. Modern combustion engines are more efficient than Lenoir's combustion engine. The original combustion engines used to mix the fuel and air together before reaching the cylinder, while modern combustion engines inject the fuel directly into the cylinder, making them more efficient.

Internal combustion engines work on the principle of using the stored power of fossil fuels to create small explosions. The power of those explosions then pushes a mechanical part that in turn rotates the automobile's wheels (by way of a series of linkages and gears).

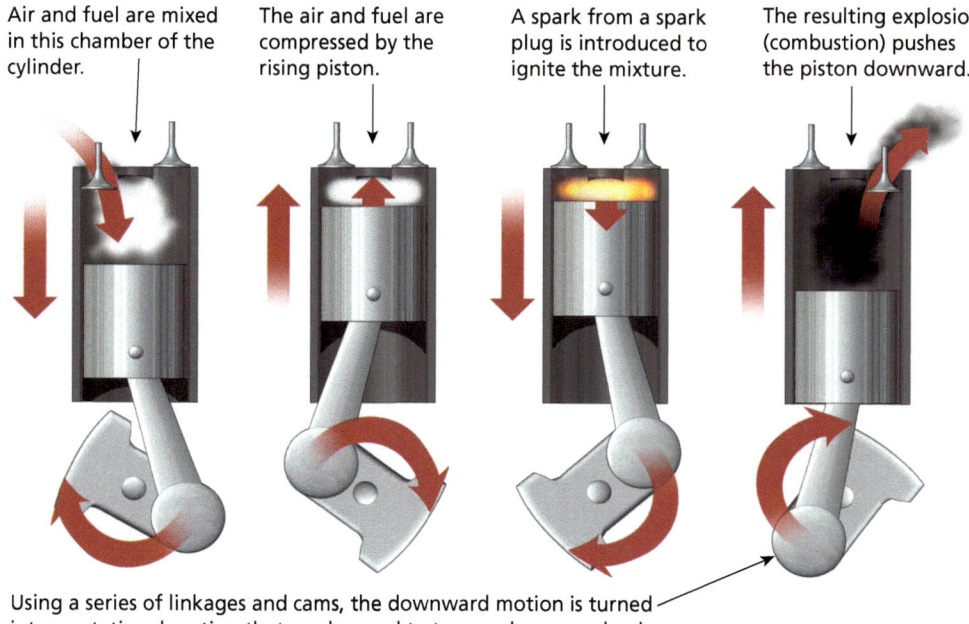

▲ A cutaway image of an internal combustion engine

The Ford Model T

The first 'affordable' car (for working-class/middle-class American families) was the Ford Model T, which was produced from 1908 to 1927. The Ford Motor Company developed one of the first mass-production assembly lines, making the product fabrication and assembly very efficient and therefore cutting down production costs. Henry Ford (owner of the Ford Motor Company) says in his book *My Life and Work* (1922):

> I will build a motor car for the great multitude. It will be large enough for the family, but small enough for the individual to run and care for. It will be constructed of the best materials, by the best men to be hired, after the simplest designs that modern engineering can devise. But it will be so low in price that no man making a good salary will be unable to own one …

He also says in the same book:

> Any customer can have a car painted any color that he wants so long as it is black.

▲ Ford Model T

Aeroplanes

Air travel has changed the world we live in and sped up the transfer of information, people and goods. In 1903, the Wright brothers achieved the first successful sustained flight and ever since then, air travel has been improving as technology advances from the development of jet propulsion to unmanned drone flights.

In the early 1900s it would take nearly ten days to reach North America from the UK on a ship travelling across the Atlantic Ocean. Until the early 2000s, you could complete the same trip in under four hours if you could afford to fly on the supersonic passenger jet Concorde. (This aeroplane was decommissioned in 2003.)

The Spitfire

▲ A Spitfire

The Spitfire was created as a short-range aircraft that could intercept and shoot down German bombers as well as protect outgoing bombing missions. Its design was hugely successful and, along with the Hurricane aircraft, the Spitfire has been credited with having been the main reason for success in the Battle of Britain in 1940. It was designed by a team of mechanical engineers headed up by R. J. Mitchell, Chief Designer at Supermarine Aviation Works. One of the reasons it was so successful was its very thin wing design that allowed much higher top speeds than other fighter planes of the time. Various design teams continued to work on variations of the Spitfire until the introduction of jet propulsion in the late 1930s.

Jet propulsion

Jet propulsion had been played around with for hundreds of years before Frank Whittle, a Royal Air Force engineer, submitted plans to his superiors and the Patent Office in 1930 for a jet-propulsion engine that could work in aeroplanes. There are different types of modern jet engine that are used in different aircraft which incorporate lots of engineering innovations. The simplest form of jet engine is the pulsejet engine which works in the following way:

- Air is sucked in through the **air intake** where it eventually meets the **fuel injector** area and mixes with the fuel.
- A spark is then added to the mix with the **spark plug** to create a controlled explosion (combustion). The resultant gases are expelled through the **exhaust**, which creates **thrust**.
- Thrust is the force that propels the aircraft through the skies.

The following diagram shows how jet propulsion works with a pulsejet engine.

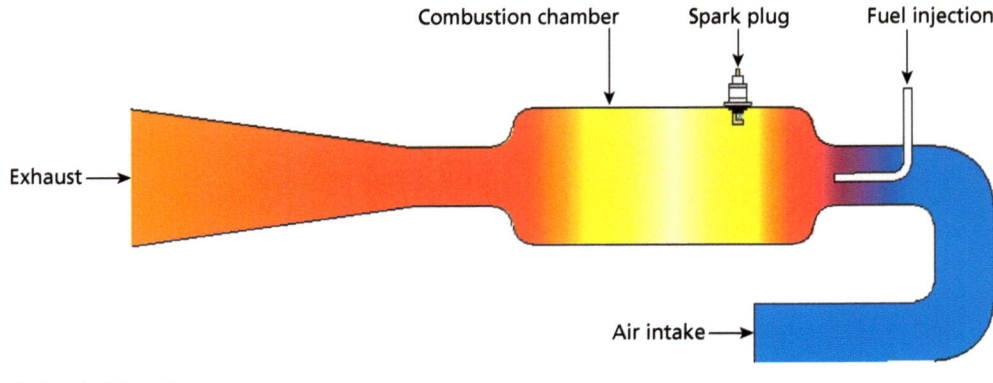

▲ A pulsejet engine

> ## Task 7.2
> Answer the following questions:
> 1. What modern materials could you use for an engine block in a car?
> 2. What properties do you need in a material used for building a car engine? Why do the materials need these properties?
> 3. Who designed the first modern internal combustion engine?
> 4. What was the first mass-produced car?
> 5. Who designed the WWII Spitfire?
> 6. What helped to make the Spitfire one of the fastest fighter planes of the time?
> 7. What force created by jets do modern aeroplanes need to fly through the sky?

7 Understanding the effects of engineering achievements

Electronic engineering

From early experiments trying to harness the power of lightning to creating storable electricity with **Leyden jars** (by Dutch scientist Pieter van Musschenbroek of Leiden (Leyden) in 1745–46), scientists have long understood the potential usefulness of electricity and the power it supplies. In the modern world, scientists and engineers have developed equipment and tools that use electricity as a power source, and innovations in electronic equipment over the last few decades have changed the face of the planet and the way humans live their everyday lives. From mining new mineral elements to global communication, the introduction of processors (microchips), computers and the internet has changed the way society functions. Think of all the electronic equipment you interact with every day and then imagine a world in which engineers have not yet discovered electricity as a power source.

Modern electronic engineering deals with the ability to design and develop controllable electronic systems that are designed to perform specific tasks. The figure below gives an example of a basic system.

▲ A basic system with an input and an output

> **Key Term**
>
> **Leyden jar** Also known as a Leiden jar, this is a simple glass jar with a foil-lined interior and exterior, used to store electric charge, much like a battery.
>
> **Semiconductor** Materials can be divided into three categories related to their ability to conduct electricity:
> - high conductivity = conductor (e.g. metals)
> - intermediate conductivity = semiconductor (e.g. silicon)
> - low conductivity = insulator (e.g. plastics).

Here you can see that an electronic system would have an input and an output. The input would be where you control the system and the output would be the required outcome. The following figure takes the idea further using the example of a hairdryer (a simple electronic system).

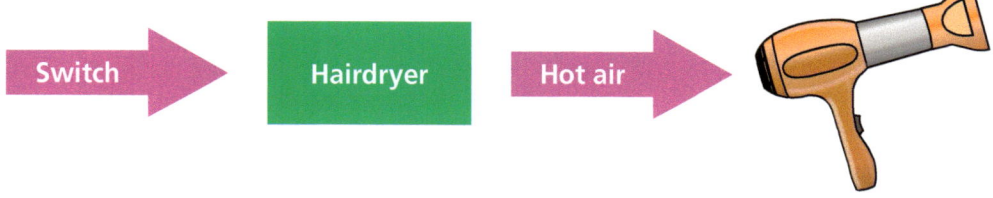

▲ The desired output from a hairdryer is hot air.

In the hairdryer example, the input would be a switch. With the switch, the electronic system can be controlled. The desired output would be the hot air; exactly what is required from the 'hairdryer system'.

Electronic engineers now deal with very complicated systems that can perform multiple tasks. Think about your mobile phone. How many tasks does it perform? Think not just about the ability to speak to friends, take photos and listen to music, but also about your phone's ability to use light in different ways, how it applies colour, tint and tone, as well as sensing the touch of your fingertips.

The transistor

It could be argued that the birth of 'modern' electronic engineering began with the discovery of a simple component called a **transistor**. A transistor is a **semiconductor** device that can be used to either boost or amplify electronic power or switch electronic signals. Prior to the invention of transistors, electronic switches and amplifiers came in the form of vacuum tubes and other large components that made any electronic device big, cumbersome and not very energy efficient.

▲ A smart watch is another example of an intricate system that can perform many functions.

▲ Augmented reality gaming with a virtual reality headset

153

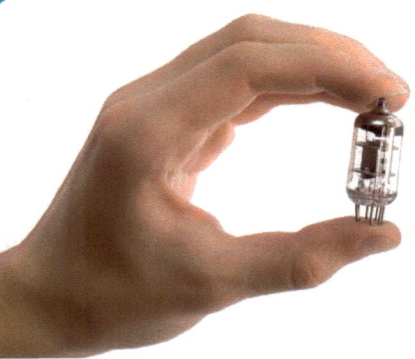

▲ An example of a small vacuum tube

With transistors, electronic devices were able to become smaller and more efficient, which allowed electronic engineers to create the powerful, mobile devices that are carried today. The first practical, usable transistor was created by the American physicists Bardeen, Brattain and Shockley, who went on to share the Nobel Prize in Physics in 1956.

The following figures help to illustrate how transistors work.

The first figure is a simple circuit with no transistor. To run power to the LED you need to control the main power unit (the battery) with a switch. This means the circuit would have a high operational cost.

The next figure shows the same circuit with a transistor. The LED will not light up as the transistor is acting as an insulator and the circuit is not complete. However, when you apply a small fraction of the voltage of the main battery (e.g. 0.8 volts), the transistor turns into a conductor and allows the main power from the battery (9 volts) to run through the circuit. This means you can operate the circuit with a much lower voltage, making the system cheaper and safer.

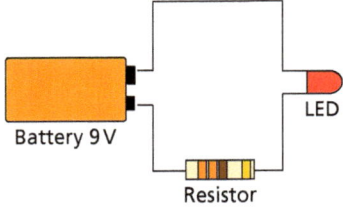

▲ A simple circuit

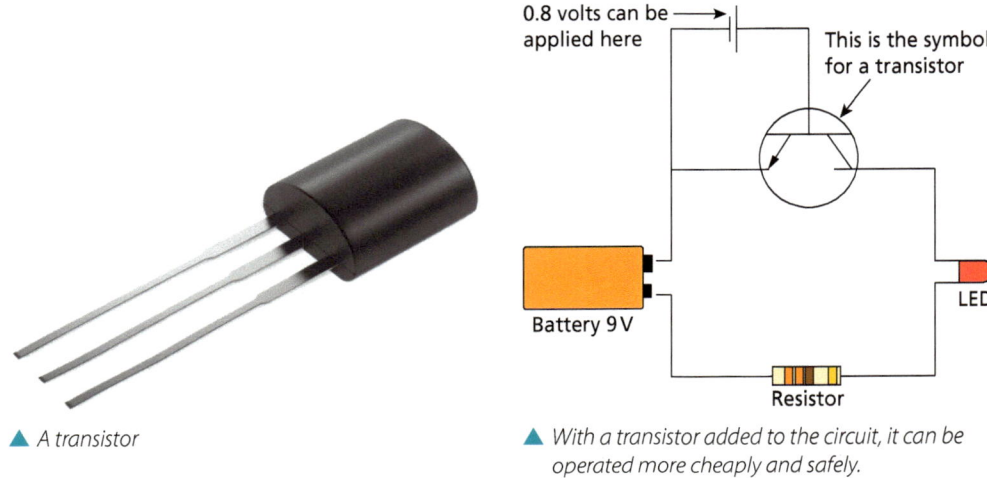

▲ A transistor

▲ With a transistor added to the circuit, it can be operated more cheaply and safely.

The transistor radio

When the transistor was introduced, it allowed electronic products to become smaller and more portable. Many designers took the opportunity to develop small, battery-operated transistor radios (radio being the main form of news and entertainment at that time). In the 1960s, the transistor radio was the biggest selling product in the Western world, allowing teenagers to listen to music away from home and people to get regular news updates wherever they were.

▲ A keyboard for a computer

The iPhone

It was the late Apple CEO Steve Jobs who initially conceived the idea of a screen that could be interacted with by way of touch for the next generation of Apple products in 2005. At the time, physical keyboards were still needed to input data into computers and phones, so designs were very limited.

A team of electronic engineers was recruited to solve the problem of 'touch-screen' and investigate the viability of it working. They developed a prototype and Steve Jobs immediately looked at introducing the new technology into mobile phones. The project was called 'Project Purple 2'.

How you interact with electronic devices and the way products are designed were changed by the team of engineers responsible for developing this new technology.

7 Understanding the effects of engineering achievements

Task 7.3

Answer the following questions:
1. The development of which electronic component changed the way electronic products were designed?
2. Describe what a 'semiconductor' is.
3. What was the best-selling electronic consumer product of the 1960s?
4. Who was responsible for changing the way we use mobile phones?
5. What was his most influential innovation?
6. What company did he work for?

Task 7.4

Describe how mobile phones have improved and developed over time. Look at:
- the design
- the materials
- the functions.

For example, screens have changed from black-and-white pixels to HD coloured screens. This has made phones more like TVs/PCs, which has resulted in more battery power being used when the phone is on.

Task 7.5

Pick one household item that makes use of technology. In your notebook, answer the following questions about your item. You can answer in bullet points, but each point must contain an explanation.
- What product/item have you chosen?
- Where is the technology in your item?
- Why was there a need for this technology?
- How has the technology advanced for your item?
- Why was this product invented? Was there a need for the product?
- How is your product manufactured?

Task 7.6

Pick one industry item that makes use of technology. In your notebook, answer the following questions about your item. You can answer in bullet points, but each point must contain an explanation.
- What product/item have you chosen?
- Where is the technology in your item?
- Why was there a need for this technology?
- How has the technology advanced for your item?
- Why was this product invented? Was there a need for the product?
- How is your product manufactured?

Task 7.7

Pick one social item that makes use of technology. In your notebook, answer the following questions about your item. You can answer in bullet points, but each point must contain an explanation.
- What product/item have you chosen?
- Where is the technology in your item?
- Why was there a need for this technology?
- How has the technology advanced for your item?
- Why was this product invented? Was there a need for the product?
- How is your product manufactured?

Explaining how environmental issues affect engineering applications

Introduction

Humans have been changing the face of the landscape for many years: from the first ancient 'engineers' who cut down trees to build wooden homes, right up to modern engineers tunnelling through seabeds to create routes for rail transport. One of the biggest eras of change, when engineers had the most impact on the world and its environment, was the **Industrial Revolution** (approximately 1760–1840). The Industrial Revolution saw a change from mainly hand-crafted products and an agrarian (farming) culture to a society where machines ruled and coal was mined to run all the newly invented steam-powered engines (used in ships, rail and industry). Factories, mass production, canals, roads and cities were built with the help of the new machines and the burning of **fossil fuels**, creating waste and pollution that were new parts of the world's environment.

In the modern era, we now have a much greater understanding of the impact engineering can have on the environment, and modern engineering looks to create and develop new solutions that either minimise the impact of engineering on the environment or even improve the environment through new innovations. In this chapter you will look at some of the areas and innovations engineers are developing that have a direct positive effect on the world's environments.

Renewable energy

Prior to the development of technologies that allowed us to harness the power of **renewable energies**, societies across the world only burned fossil fuels to create power. In fact, fossil fuels are still the main source of power for the world today, with oil being the resource that is predominantly used to power industrial machines and transport, and create materials such as plastics. The problem with using fossil fuels (a **non-renewable** resource) is that that they not only cause harm to the environment by the pollutants they create when burnt, but they are also a finite resource and will eventually run out.

Due to the huge drawbacks of using fossil fuels, engineers have been looking at creating technologies that harness renewable energy sources such as wind, solar, hydro and geothermal.

Wind power

Wind power or wind energy uses wind turbines to capture the energy from the movement of air. The turbines use propellers to turn a generator to create electrical energy that is then either stored in batteries or fed directly into the electric grid. Wind farms are dedicated areas where lots of wind turbines can be grouped together to capture the energy of air currents. Wind turbines can be expensive – their mechanical nature means they need to be maintained; however, they have minimal impact on the environment and make good use of a renewable energy source.

Solar power

Solar power is the ability to catch and convert sunlight into electricity that is either stored in batteries or fed directly into a power grid. Sunlight is captured and converted using solar panels, which are now quite a common product and can be seen adorning the rooftops of many houses. Solar panels use photovoltaic cells to convert sunlight into electricity by allowing **photons** to knock electrons free from atoms which then create a flow of electricity. The largest solar farm in the world is the Bhadla Solar Park, located in the Thar Desert of Rajasthan, India. It covers an area of 56 square kilometres.

>  **Key Terms**
>
> **Fossil fuels** Non-renewable resources, including coal, oil and natural gas, that can be burned to release energy.
>
> **Renewable energies** Sources of energy that can be renewed naturally such as wind, solar, geothermal and tidal.
>
> **Non-renewable** When describing a source of energy, it means that the source is finite and will eventually run out.
>
> **Photons** Light particles.

▲ Wind turbines on a hillside capturing the energy of the movement of air

▲ A solar farm

To run solar panels efficiently, however, you need to have good air quality to ensure the sunlight can actually reach the solar panels. China (one of the world's largest manufacturing economies) still burns huge amounts of fossil fuels for its industries, alongside the use of newer technologies such as solar panels. The continued use of fossil fuels is causing a problem for the new technologies, as the air quality is so poor in some areas that the smog and air pollution block the sunlight from reaching the solar panels.

Hydropower

Hydropower involves harnessing the power of moving bodies of water (e.g. tides) to turn turbines to generate electricity which is then either stored in batteries or fed directly into power grids.

Water has been used to power machines for centuries, with early watermills using the power of the flow of a river to turn a wheel. This, in turn, would turn grindstones for grinding grain (see page 150). On a large scale, hydropower can be obtained in several ways, from building huge dams and reservoirs as part of hydroelectric plants, to creating floating turbines out at sea to harness the power of the waves. A great advantage in using water to generate electricity is that it is easy to predict the tides, the flow of a river or even when you open the sluice gates of a dam to allow water to flow. This predictability makes hydropower a potentially very efficient source of renewable energy.

▲ *The power of the river flow turns the water wheel – an early example of hydropower.*

Geothermal power

Geothermal power is the harnessing of the Earth's natural warmth to raise the temperature of water and create steam. The steam then turns turbines which generate electricity. This is a clean and renewable resource that is currently being used by many countries in the world including Kenya, Iceland and New Zealand. Currently, Iceland is able to generate over 25% of its electricity from geothermal sources. Although geothermal energy has many benefits to the environment, it is expensive to create geothermal plants and difficult to find locations in the world where they can be built.

The product lifecycle

When creating new innovations and solutions, engineers need to understand the impact their design is going to have on the planet. Modern engineers must ask: what will it be made from, where will I obtain the resources needed, how will it be used and what will happen to it when it comes to the end of its life? By addressing these questions, modern engineers can produce products and solutions that may minimise their solutions' impact on the world's environment. The responsibility of using the world's resources lies with the decisions made by the engineer when starting a project. They must make choices that ensure there is minimal damage to the environment.

Understanding these issues has led to the development of the product lifecycle assessment. This model looks at the overall impact of the creation of a new product and allows decisions to be made at each stage that could potentially minimise the impact on the environment.

The engineering design process can have a major impact on each of the processes listed in the figure. Engineers, therefore, have a huge responsibility to look after the environment and need to consider design choices that would affect each of the above areas. Can you choose a better material? Can you use less transport? Could the product be made reusable or recyclable? Can you extend the usable life of the product? Can you find a way to prevent the decline in sales that tends to happen over time (see figure below)?

End of product life
What happens to the product at the end of its life? Can any or all of it be recycled/reused? Is it easy to dismantle and recycle? Will it go into landfill (rubbish dump)? How can this now useless product have a minimal impact on the environment?

Using the product
How will the product be used? Have you created it to only last a limited time? Is it an optimal design that will last a long time? Does it need servicing or maintenance? Are there extra environmental costs if the customer uses it (e.g. power usage)?

Assembling parts
How is your product going to be assembled and packaged? Will your packaging use even more material? Can it be assembled in the same place as it was manufactured? Once assembled, does it need to be transported to the customers?

Extracting raw materials
What and where are the materials you need for your solution? Do you have to transport them from the other side of the world? Can you source them locally? How are you going to extract them? Can you find recycled materials to use?

Refining raw materials
Do the materials you have specified need refining (e.g. crude oil into plastic)? How many refining processes are your materials going to need? Can you substitute some raw materials for recycled/reused materials that have already been refined?

Manufacturing parts
Where is your product going to be manufactured? Does it need to be transported when made? Do you need to set up a new manufacturing plant and train new staff? Can you manufacture locally?

▲ Product lifecycle assessment

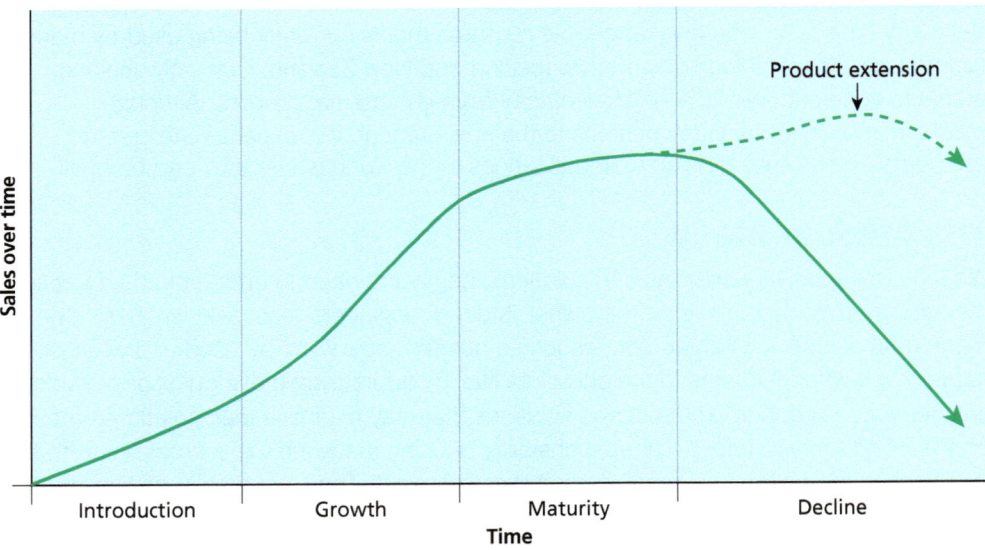

▲ An overview of the product lifecycle

Design for maintenance
Maintenance means any activity which allows the product to have a longer life. It can include anything from being able to repair worn out parts to allowing the replacement of batteries. Designing a product to allow maintenance may mean including features such as access panels and standard screws. These make it easier for parts to be replaced. Alternatively, products might be made from a series of standard modules. This would mean that if the product went wrong, only the faulty module would need to be repaired or replaced. Using modular design also makes it easier to upgrade and improve products as there are new developments and improvements in technology.

Manufacturing

Converting raw materials into finished products has a significant impact on the environment. By analysing existing manufacturing processes it is possible to identify areas that can be modified to achieve cleaner and more efficient processes.

Companies can lower their environmental impact and production costs by reducing the amount of material used. This could involve using less packaging, reducing waste and consuming less energy during manufacture. As a direct result of such action, the environmental impact of manufacture is significantly reduced.

Modifying design

One of the best ways to reduce the environmental impact of a product is by modifying the design of the item and/or its packaging. You may be able to improve the product's efficiency by:

- using a simpler design with fewer components and so reduce materials and assembly time
- using lighter materials to reduce their weight or using less of each material
- using materials that use less energy during manufacture and produce less waste
- using simpler components that are easier to machine or mould
- using a simpler or different workflow with improved quality control.

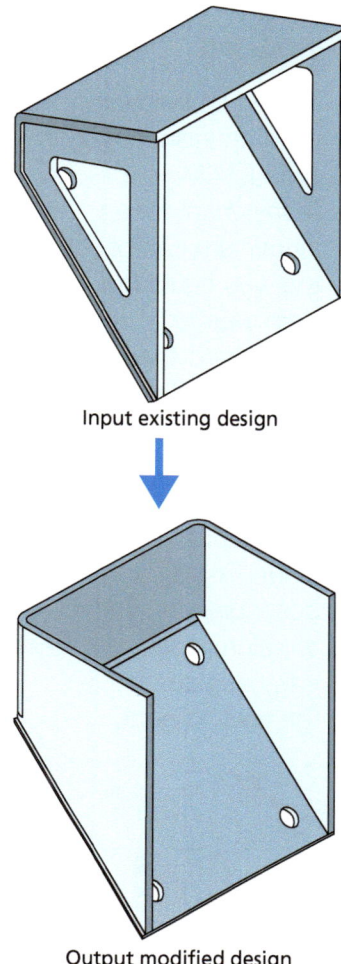

Input existing design

Output modified design

▲ Designs can be modified to reduce their environmental impact.

Distribution

The UK has identified a number of issues relating to the cleaner distribution of goods, but they all result in the same key concerns: extremely large energy use resulting in high carbon dioxide emissions, which contribute towards global warming. The carbon footprint of a product can be reduced by:

- reducing congestion on the roads – companies are under pressure to consider more environmentally friendly methods such as trains (particularly electric trains) and waterways where possible
- shortening journeys – companies can buy materials from local sources and use distribution centres to prevent wasted mileage
- saving fuel – companies can reduce the weight or size of products to enable more to fit in one lorry.

Use

Making a product uses resources, such as raw materials and energy. This has an impact on the environment. There are a number of things that an engineer might think about to reduce environmental impact:

- the material used to make the product
- the life of the product
- what happens to the product at the end of its life
- how the product can be used.

Disposal

At the end of their useful life, most products are disposed of in some way. How this is carried out can have a significant impact on the environment. A large proportion of the products we use currently ends up in landfill; this means that the products are buried in underground rubbish dumps. This is one of the least environmentally friendly methods of disposal.

Task 7.8

The figure below is an unfinished product lifecycle assessment for an aluminium can of cola that you could buy from your local shop. Copy the diagram into your notebook and complete each stage of the product lifecycle assessment for this product. Write a short paragraph describing the impact the aluminium can has had on the environment.

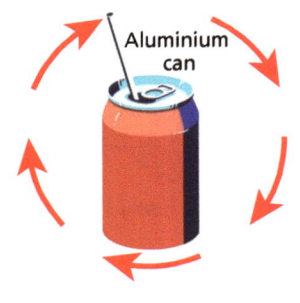

How can engineers aid the disposal of an item?
- Directives – including **WEEE** (Waste Electrical and Electronic Equipment recycling) – help to ensure that when a product is finished with, the elements can be recycled.
- Use higher quality materials to make products. If they are less likely to break, there will be less waste.
- Design products in such a way that they are easier to fix with parts available for this to happen.
- Recycle – use the product for another item or recycle parts to create a new item.

Biodegradability

If a product has to be disposed of in landfill, the impact on the environment will be a lot less if the materials from which it is made are **biodegradable**. Many modern products are made of plastic, derived from crude oil. The majority of these plastics are not biodegradable and some may take hundreds of years to break down naturally. While there may be limited options for the product itself to use biodegradable materials, this is an especially important factor when considering packaging materials, which are usually thrown way without any further use.

Cost

An engineer must consider the following costs when designing a product:
- production costs
- breakdown costs
- transport costs
- disposal costs
- costs to use the product
- maintenance costs
- repair costs.

Social responsibility

Social responsibility means ensuring that people's quality of life and human rights are not compromised to fulfil the expectations and demands of the client when developing new products.

Economic responsibility

Economic responsibility is about considering the financial implications of our actions. Engineers should work to ensure that there is an economic benefit both to the region from which the product came and to the region in which it is marketed.

Environmental responsibility

Environmental responsibility is about ensuring that our actions and lifestyles don't cause the planet's resources to be used at unsustainable rates.

Existing and future engineering materials and processes

Engineers and scientists around the world are constantly looking to develop new materials and engineering processes that are not only more efficient and cost effective but are also better for the environment and sustainable in the long term. Whether it is reducing the amount of raw materials used by substituting recyclable materials or developing new and exciting ways of building and manufacturing, engineers should now prioritise the environment when creating new solutions and incorporate new technological advancements when making decisions. Here are a few examples of some innovative practices that are currently being used or will soon be used in product and engineering developments.

- **Sustainable concrete:** Concrete is an excellent building material made of cement, air, water, sand and gravel. It is possible to use items such as crushed glass, wood chips and slag to add bulk to a concrete mix and so use lower amounts of raw materials.

Key Term

Slag The waste material that is left when smelting or refining metals from their ores.

- **Pollution-absorbing bricks:** These bricks can act as air filters for the immediate environment around them. They absorb fine to coarse particles from the exterior air and so reduce air pollution inside a building.
- **Bioplastic:** Instead of using chemicals from crude oil, bioplastics derive from organic matter and plants such as sugar cane, algae, corn starch and crustaceans. Apart from not using non-renewable fossil fuels, bioplastics are also 100% biodegradable, ensuring there are no harmful effects to the environment when disposed of.
- **Photovoltaic surfaces:** Much like solar panels, photovoltaic surfaces are an advanced use of solar-power technology but, unlike solar panels, they can be applied to surfaces such as glass. Large structures such as skyscrapers could be self-powered and self-sustainable if covered in a photovoltaic film.
- **Self-healing material:** Although in the early stages of development, this is a material that has the potential to 'self-heal'. By using the carbon in the atmosphere, self-healing materials could repair themselves when broken.
- **Smart factories:** Many factories employ computer-integrated manufacturing and robotics, but smart factories are the next level. Smart factories are able to adapt and change processes through real-time monitoring, as every aspect of the manufacturing process is managed by digital computers and robotics. In a smart factory, there would be almost no human input in the day-to-day manufacturing of products. This would be good news in terms of efficiency but bad news in terms of employment opportunities.

▲ *An engineer uses a tablet to monitor a robot arm in a smart factory.*

What is sustainability/sustainable engineering?

In terms of engineering and creating new products, **sustainability** means developing products that:
- are made from sustainable resources
- use minimal resources/renewable resources during manufacture and transport
- can be recycled fully.

As an example, a team of structural engineers sets out to create a sustainable office building. To achieve this goal they may have to:
- source materials from a renewable resource
- use materials that are 100% recyclable
- minimise the use of any non-sustainable materials
- source materials locally
- create a building design that has a minimum impact on the environment
- ensure the building operates efficiently in terms of energy use (renewable energy sources)
- ensure there is minimal maintenance needed for the building
- ensure the building is easy to renovate and update if needed
- ensure the building is easy to dismantle, demolish and recycle at the end of its life.

Sustainable engineering not only minimises the impact of creating new products by not using 'virgin' resources but can also have a positive impact on environments.

Key Term

Sustainability In general terms, the ability to maintain or sustain something at a particular level. In terms of engineering and product design, sustainability relates to the creation of products with a minimal impact on the environment.

The 6Rs of sustainability

Many consumers are trying to think of the environment and sustainability when they buy things; they are thinking of 'green' issues. Designers and manufacturers are required by law to try to reduce the environmental impact of the products they create. The 6Rs summarise some approaches that can be taken by the designer and manufacturer. The 6Rs are:
- recycle
- reuse
- repair
- rethink
- refuse
- reduce.

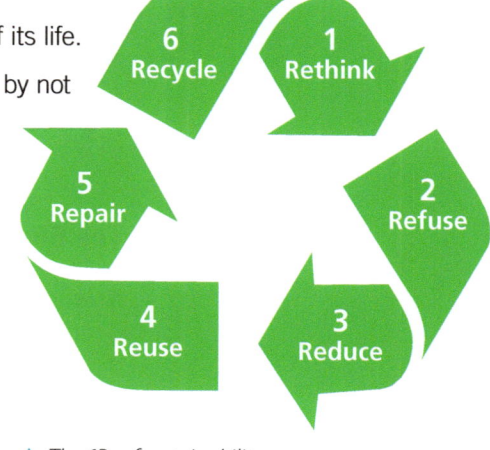

▲ *The 6Rs of sustainability*

Recycle

Products are converted back to their basic materials and remade into new products. Examples include:

- glass crushed, melted and made into new bottles
- aluminium cans melted down to make new products
- plastic bottles recycled into drainage pipes and clothing.

Designers and manufacturers of products need to design products for recycling.

▲ Milk bottles are collected and refilled – an example of reuse.

Reuse

Glass milk bottles are a classic product that is reused. A more recent product that can be reused is a printer cartridge, which can be refilled. Some products now have filters that can be washed rather than using disposable, single-use filters. Designers need to consider how a product may be dismantled at the end of its life so that parts may be reused.

Repair

Designers have a responsibility to design products that can be repaired more easily. It takes fewer resources to replace a part of a product than to replace the whole item. Some electronic companies make it difficult for you to replace a part easily if one of their products breaks. This is called **planned** (or built-in) **obsolescence**.

▲ Designers need to ensure that new products can be repaired.

Rethink

By rethinking design and materials, engineers can make products that do the same job more efficiently. For example, electric cars have exceptional environmental performance compared to traditional internal combustion models, operating solely on electricity and eliminating the need for petrol or diesel. In the mobile phone industry, allowing users to easily replace a device's battery when needed extends the product's lifespan.

Engineers can also design packaging so that it is easier to recycle, for example, by making it from a single material.

Refuse

Increasingly, engineers need to start thinking about how the consumer will react to their products. Will they refuse to buy them? Are they made from a material that is harming the environment? Consumers will make informed choices about the products they buy.

Reduce

Engineers are looking to design products that:

- use fewer materials in the manufacture
- take less energy to manufacture
- need less packaging during transport.

Retailers can reduce carbon emissions by transporting products straight to the consumer from the place of manufacture, instead of via warehouses and shops.

The 6Rs and global companies

Nike

Sports brand Nike has developed a scheme to make use of the material from old, discarded Nike trainers. Through collection points, Nike collects unwanted trainers and uses them in a material it calls Nike Grind. The raw materials taken from the athletic trainers include: rubber from the outsole, foam from the midsole and fabric from the upper. These materials are ground up and used by companies in sport and playground surfaces.

7 Understanding the effects of engineering achievements

Cadbury
In recent years, Cadbury (owned by Mondelēz International) has changed the packaging it uses on its Easter eggs. It removed the plastic windows from all its chocolate shell eggs in 2021. It also moved to 100% sustainably sourced cardboard and removed 108 tonnes of cardboard from the market.

ASDA
ASDA was the first major UK supermarket to stop selling 5p plastic carrier bags in its stores in 2018. The move saw the sale of reusable bags increase by 1200%. The initiative is part of a plan to help the retailer reduce the environmental impact of plastic bags. Bags for life are more durable and long lasting than regular carrier bags. ASDA also encourages customers to bring their own bags.

IKEA
Wood is one of IKEA's most important manufacturing materials; it is used in many of its products. Over recent years, IKEA has been increasing its use of wood accredited by the FSC (Forest Stewardship Council). In 2022, 99.9% of the wood used by IKEA was either recycled or came from FSC-certified forests.

Recycling materials
Millions of tons of materials are used every year to produce products from plastic packaging to new buildings, with a lot of the materials eventually ending up in landfill sites or being thrown into the oceans as waste instead of being repurposed or recycled. The new goal for engineering and manufacturing is to create a material-neutral world, where all the resources we extract could either be used again and again or would fully biodegrade into a non-harmful (to the environment) substance. Clearly, we are not there yet but we are getting closer as innovations and technology improve. A good example would be replacing plastics derived from crude oil with 100% biodegradable bioplastics.

The recycling of materials is now written into law, with the ISO (International Organization for Standardization) setting world standards for the percentage of materials that have to be recyclable when new products are manufactured. For example, the standardisation number ISO 15270:2008 deals with the percentage of plastics that must be recyclable for new products. All manufacturing companies have to conform to this international law.

Emblems and labelling for recycling and sustainability
When you use or purchase a new product, you may see logos printed or stamped onto the packaging or surface, showing whether the product comes from a sustainable source or whether parts of it can be recycled.

There are three types/groups of recycling:
- **primary**
- **secondary**
- **tertiary**.

Primary recycling is when you reuse something in the same way. Examples include giving clothing to a charity shop, selling an item at a car-boot sale or taking clothes to a clothes bank.

Secondary recycling is when an old product is converted into another product or repurposed into something new. You can still tell that the product consists of the old one. Examples include knitting old fabric strips into a scarf or making a patchwork quilt out of old curtains.

Tertiary recycling is when a totally new product is formed out of an old one. After tertiary recycling, you are not able to tell what the product originally was. Examples include bags made out of plastic bottles or flooring made from old tyres.

▲ *The recycling logo (mobius loop) shows whether a product or part of a product can be recycled.*

163

▲ Most plastics are recyclable.

Recycling plastics

Most plastics can now be recycled. However, the recycling process for plastics can be difficult and does tend to use a lot of energy. In addition, plastic degrades every time it is recycled and so can only be recycled a limited number of times. All plastic products must now be stamped with details of the type of plastic they are made from so they are easier to recycle. This should reduce the costs involved in making new plastics.

Recycling is an expensive process, so any well-engineered design that is easier to recycle will reduce the cost to society and the environment. Modern engineers now work hard to create solutions and products that consumers and industry find easy to recycle.

The following figure is a plastic recycling chart that explains what the 'stamped' numbers on plastic products mean and what materials they are made from.

1 PET	2 HDPE	3 PVC	4 LDPE	5 PP	6 PS	7 OTHER
polyethylene terephtalate	high-density polyethylene	polyvinyl chloride	low-density polyethylene	polypropylene	polystyrene	
soft drinks bottles juice containers cooking oil bottles	milk bottles cleaning agents laundry detergents shampoo bottles	chocolate box trays clear plastic packaging bubble wrap food packaging	shopping bags squeezable bottles plastic sacks	toys luggage car bumpers plastic chairs	toys hard packaging items CD cases	other plastics such as: fibreglass acrylic nylon

▲ A recycling chart for plastics

Recycling metals

Metals are considered a sustainable material as they are almost 100% recyclable and reusable. (An exception is the loss resulting from the corrosion (rust) of ferrous metals.) As a result, using metals in products and construction is generally good news. In contrast to plastics, metals can be recycled again and again. This is good news for engineers and the environment. One of the most common metals used today is steel, which is probably the most recycled of all metals. A steel car you drive around in today could have previously been a washing machine.

Task 7.9

Using your new knowledge of engineering and the environment, answer the following questions:
1. Name three renewable energy sources.
2. Explain the term 'sustainability'.
3. What building materials could be considered sustainable and why?
4. Which organisation sets standards on recycling targets for new products?
5. What recycling logo could be displayed on new products and what benefits would this bring to the consumer?

Task 7.10

In engineering, the environment and recycling are vitally important when designing a product. Copy the image below into your notebook and add annotations to identify the parts of the car that can be recycled. Provide details of what they can be recycled into.

Understanding properties of engineering materials

8

In this chapter you are going to:
- learn about properties of materials
- understand the different categories of materials
- identify materials of named products
- explain why materials were chosen for specific tasks
- understand and name specific smart materials.

This chapter will cover the following areas of the WJEC specification:

Unit 3 Solving engineering problems: 3.2 Understanding properties of engineering materials		
• 3.2.1 Understanding materials, their properties, and their selection for specific purposes	• 3.2.2 Describe properties required of materials for engineering products	• 3.2.3 Explaining how materials are tested for properties

Understanding materials, their properties and their selection for specific purposes

Introduction

In this section you are going to find out all about different materials and what some of those materials can do. Materials are the fundamental make-up of the world we live in so, as an engineer, it is your job to discover what the capabilities are for each of the materials you intend to use. When we talk about materials we are really talking about properties.

So what are material properties?

The properties of a material explain to us exactly what the material does; in other words, what that material is good at doing and what it is bad at doing. Once we know what materials can or cannot do then we can understand how and where they can be used and begin selecting the right ones to perform different tasks.

For example, when asked to create a structure that allows traffic to cross a river or gorge, an engineer would first need to discover what properties would be required in order for the structure to be strong and rigid enough to span the gap and support the traffic. Once those properties were identified then they would pick the appropriate material.

In Ancient Rome, engineers were masters of understanding materials and properties. By using their knowledge effectively, they were able to quickly build bridges that allowed them to march their armies all over Europe, conquering the known world. This gave the Romans a huge advantage as, without this knowledge, other cultures were severely restricted in their movements.

Properties of materials

The property of a material dictates how it will perform and react to the environment it is functioning in and how it will react to the job you have asked it to do.

The following table details some properties and their definitions. In the table, specific names or words are used to identify a property. Learn these words and their meanings, as engineers have to describe material properties all the time for each project they undertake.

Property	Definition
Elasticity	Ability to regain its original shape (e.g. rubber)
Ductility	Ability to be stretched without breaking and drawn into wires (e.g. copper)
Malleability	Ability to be pressed, spread out or hammered into new shapes (e.g. lead)
Hardness	Ability to resist scratching, cutting or wear and tear (e.g. high-carbon steel)
Work hardening	Property changes due to working (e.g. bending steel back and forth)
Brittleness	Will snap easily and will not bend (e.g. glass)
Toughness	Resistant to breaking and bending (e.g. cast iron or urea formaldehyde polymer)
Tensile strength	Retains strength when stretched (e.g. some aluminium alloys)
Compressive strength	Very strong under pressure (e.g. concrete)
Corrosive resistance	Ability to resist corrosion in the environment in which it is working (e.g. iron rusts if not treated)
Conductivity (electrical)	Ability to conduct (transmit) electrical current (e.g. copper wire)
Conductivity (thermal)	Ability to conduct heat (most metals such as steel cooking pans)
Environmental degradation	How the material corrodes and degrades in an environment (salt water, weather, fire)

Key Terms

Urea formaldehyde polymer A hard, slightly brittle plastic used for electrical casing/housing.

Environmental Relating to or caused by the surroundings in which someone lives or something exists.

Degradation The process of changing to a worse condition.

Task 8.1

Below are some statements relating to material properties. Copy them into your notebook and have a go at filling in the blanks.
1. We use cast iron for manhole covers because it is _____ and therefore is unlikely to bend or break when traffic drives over it.
2. We have some very nice copper-bottomed pots and pans at home for cooking and they are great at _____ heat from the stove.
3. Don't throw stones near glass windows. They are _____ and likely to break.
4. My drill bits are made from _____, which makes them very _____ and good for cutting through other metals.
5. Most modern bridges are made from _____ and _____, as they have great _____ and _____ properties and can therefore bear the weight of all the traffic.

Materials

Now that we understand about the properties of materials, we can begin to identify the materials themselves, the categories they fall under and the functions they can perform (depending on the properties they have).

In this section we are going to look at:
- metals
- plastics
- composite materials
- smart materials.

Metals

A large range of metals exist in the world, all with different, useful properties and all used to perform different tasks, from the gold and copper used to produce your mobile phones to the steel and aluminium used to create the skyscrapers seen in the world's largest cities. Metals generally fit into two categories:
- **ferrous** metals
- **non-ferrous** metals.

There is a sub-category: **alloys**. These are mixtures made from ferrous and non-ferrous metals and sometimes other elements. Steel, brass and bronze are alloys.

Ferrous metals

Iron is the fourth most common element (the second most common metal) in the Earth's crust and is therefore easy to find. Pure iron tends to be too soft to use on its own, so other metals and elements are mixed with it to create useful materials (for example steel which is made from iron mixed with carbon).

You can easily identify a ferrous metal in one of two ways. Firstly, iron corrodes (rusts), so anything with rust on the surface, as a result of a process called **oxidisation**, must have iron in it. Secondly, iron has magnetic properties and can be identified readily if you have a magnet to hand. Some metals that contain iron have been treated to have corrosive-resistant properties, however, and so may not rust (e.g. stainless steel).

Properties of ferrous metals
- They contain iron.
- They may have small amounts of other metals or elements added, to give the required properties.
- They are magnetic.
- Unless treated, they give little resistance to corrosion.
- They can be recycled again and again without losing quality, therefore scrap metal from vehicles and demolition sites, and offcuts from manufacturing industries can be readily reused.

Key Term

Oxidisation In the process of oxidisation, steel/iron surfaces react with oxygen from the atmosphere and create ferric oxides (for example, rust).

The table below lists some examples of ferrous metals with their properties, composition and uses.

Material	Properties	Common uses	Composition
Mild (low carbon) steel	• Good tensile strength • Tough • Corrodes easily	Used in many products such as: • carcasses for PCs • Xboxes, etc.	• Iron • 0.1–0.3% carbon
Medium-carbon steel	• Tough • Medium strength • Ductility	• Railway tracks • Train wheels • Gears • Machinery parts	• Iron • 0.3–0.6% carbon
High-carbon steel	• Tough • Hard • Can be brittle	Tools such as: • saw blades • drill bits	• Iron • 0.5–1.5% carbon
Stainless steel	• Resistant to corrosion (rust) • Tough	• Medical instruments • Cutlery	• Iron • Nickel • Chromium
Cast iron	• Good compressive strength	• Drain and manhole covers • Engine blocks	• Iron • 2–6% carbon

Task 8.2

1. In your notebook, write a use for each type of ferrous metal listed here:
   - cast iron
   - medium-carbon steel
   - stainless steel
   - high-carbon steel
   - mild steel.
2. Explain the term 'ferrous metal'.
3. Give an example of a suitable ferrous metal that may be used to make garden tools. Explain your reason for selecting this metal.
4. Which type of ferrous metal would be suitable to make tough kitchen sink units? The surface would have to withstand wear and tear and flooding with water.

Non-ferrous metals

Non-ferrous metals do not contain any iron. There are lots of examples of non-ferrous metals including aluminium, gold and copper. Non-ferrous metals can have different properties from ferrous metals and have a range of different uses. Copper, for example, is a great conductor of heat and electricity and so is commonly used in electrical cables/wires and cooking pans. Non-ferrous metals also tend to have much greater resistance to corrosion than ferrous metals but they do not have magnetic properties.

With the exception of aluminium (the most abundant metal in the Earth's crust), most non-ferrous metals are less common than iron. They also tend to be much more expensive to refine from their metal ores and more expensive to fabricate when compared to iron.

8 Understanding properties of engineering materials

A rotary washing line is an engineered product that performs a task in an external environment. It is likely the washing line is going to get rained on and would therefore encounter environmental degradation. You can purchase rotary washing lines that are made from aluminium (a non-ferrous metal), as they are corrosion resistant and light. However, you can also purchase washing lines made from steel (99.9% iron and 0.1% carbon) that may rust and are very heavy. So why purchase a steel washing line? Steel is a much cheaper material and is also much easier to fabricate. Steel can easily be welded with common welding equipment, whereas aluminium would need specialist welding equipment. All these factors mean that the steel washing line is much cheaper to purchase for the consumer.

▲ Rotary washing lines can be made of aluminium or steel.

Properties of non-ferrous metals
- They are metals which do not contain any iron.
- They are not magnetic.
- They are usually more resistant to corrosion than ferrous metals.
- They tend to be more expensive than ferrous metals.
- Non-ferrous metals are used because of their desirable properties such as low weight (aluminium), high conductivity (copper) or resistance to corrosion (zinc).

Uses of non-ferrous metals
The table below lists some examples of non-ferrous metals with their properties, composition and uses.

Material	Properties	Common uses	Composition
Aluminium	• Light • Soft • Malleable	• Good for alloys • External products • Aircraft	• Aluminium
Lead	• Ductile • Malleable • Heavy	• Roofing • Batteries	• Lead
Copper	• Good conductor of heat and electricity • Ductile	• Piping • Electrical wires • Pans	• Copper
Gold	• Soft • Malleable • Shiny • Tarnish/corrosion resistant	• Jewellery • High-end stereo connections	• Gold

Alloys

An alloy is a mixture of elements that usually has a metal as the major/parent component. Steel, for example, is 99.9% iron and 0.1% carbon. Alloys were developed to create materials with different properties from those of the pure metal. By heating up, melting and mixing different elements, you can create materials with new and very useful properties.

Bronze is an alloy that is created by mixing copper and tin. Bronze is harder, more corrosion resistant and easier to melt and cast into different shapes (e.g. axe heads) than both its parent metals.

The table below lists examples of some common alloys with their properties, composition and uses.

Alloy	Properties	Common uses	Composition
Duralumin	• Lightweight • Strong • Extremely corrosion resistant	• Car parts • Aircraft parts	• Aluminium • Copper • Magnesium • Manganese
Brass	• Malleable • Hard • Corrosion resistant	• Musical instruments • Ornamental products	• Copper • Zinc
Stainless steel	• Shiny • Durable • Strong • Corrosion resistant	• Medical instruments • Cutlery	• Iron • Nickel • Chromium

Alloying agents/elements

Modern engineers use many different alloys that perform a varied number of tasks. Different **alloying elements** can be added to a mixture in known ratios to confer specific properties on the alloy. The following table details some common alloying elements and the properties that they can add to an alloy.

Alloying element/agent	Properties
Nickel	• Increases strength • Increases hardness • Increases resistance to corrosion
Chromium	• Increases hardness • Increases resistance to corrosion
Vanadium	• Increases toughness of steel • Increases wear resistance

Task 8.3

1. Explain the term 'non-ferrous metal'.
2. Name a non-ferrous metal that is suitable to make ornaments?
3. What is the difference between ferrous and non-ferrous metals in terms of their properties?
4. What is an alloy?
5. Is copper an alloy? Explain your answer.
6. Describe some practical uses of brass.
7. In your notebook, match the name of the non-ferrous metal or alloy to its use.

Brass   Bronze   Pewter   Lead   Aluminium   Copper   Zinc

Plastics

Imagine a magical material that you could mould into any shape you can imagine. It could be any colour you want, would never rust, is very low cost, could have any finish applied to it (rough, gloss), is incredibly strong and very lightweight.

Welcome to plastics!

Regardless of all the negative impacts plastic has had on the world, the reason it is used so widely is because of all the amazing properties it has. It is easy to create engineered solutions to problems when you have a material that can do all these incredible things. This is why so many engineers and designers specify its use.

▲ *Plastic is an incredibly versatile material with many useful properties.*

Common properties of plastics

In general, plastics are:
- flexible – you can squeeze the shampoo out of the bottle
- watertight – the shampoo will not leak
- shatterproof – the bottle won't break if dropped
- light – easy and cheap to transport
- easily moulded – can be made into complicated shapes
- heat insulators – for example, plastic vending machine cups
- electrical insulators – plugs and sockets are made of plastic
- durable
- non-biodegradable.

Problems with plastics

All of the problems with plastics relate to what happens to them once they have been used and the issues with their safe disposal.
- As a result of plastics being non-biodegradable, they do not readily decay when thrown away.
- Plastics cannot always be recycled easily.
- The quality of a plastic decreases every time it is recycled so, unlike a metal, it cannot be recycled endlessly.
- The vast quantities of plastic used in everyday life means that large amounts of plastic end up in landfill sites.
- Some local authorities burn plastics to dispose of them which results in dangerous fumes being produced.

Where do plastics come from?

Plastics are made from the chemicals that are extracted from crude oil. Crude oil is extracted from the ground and transported to a refinery where it goes through the process of being refined. Many different chemicals are extracted from the refinery process, one of which is naphtha. Naphtha then is processed further to produce plastics using the **polymerisation** process.

Polymers is the scientific term used to describe plastics. Polymers are long, intertwined chains of molecules, rather like spaghetti. They consist of carbon and hydrogen atoms with other elements attached to give plastics their different properties.

The following diagram shows how naphtha is extracted from crude oil in an industrial process called **fractional distillation**.

> **Key Terms**
>
> **Polymerisation** The industrial process used to create plastics from naphtha. In scientific terms, polymerisation is the name of the process in which small molecules – called monomers – are joined together chemically to make a long, chain-like molecule called a polymer. Plastics are also called polymers.
>
> **Polymers** Large molecules made up of many monomers joined together. Polymer is the scientific name for a plastic. *Poly* is the Greek word for many.

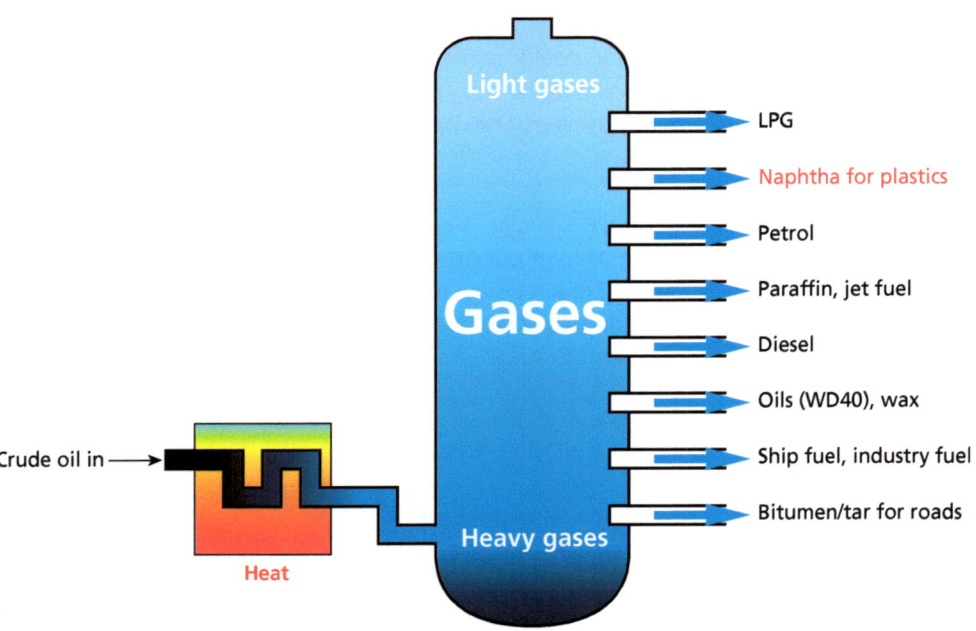

▲ The separation of crude oil into fractions, including naphtha which is then used to produce plastics.

From naphtha we derive **monomers**, which are then joined using the polymerisation process to create polymers.

▲ Polymerisation is the joining of small molecules called monomers to form a very long molecule called a polymer.

Key Term

Monomers Atoms or small simple molecules. The term comes from the Greek words *mono* = one and *meros* = part.

Types of plastics

Plastics can be separated into two main categories:
- thermoplastics
- thermosetting plastics.

Thermoplastics

These plastics can be reshaped when heated and so can be remoulded. They become mouldable on heating as they do not undergo significant chemical change. Reheating and shaping can be repeated. The bonds between the molecules are weak and become weaker when reheated. In addition, there are no cross-links between the chains of monomers. Reheating can be done multiple times before the thermoplastic starts to degrade. Thermoplastics are also recyclable.

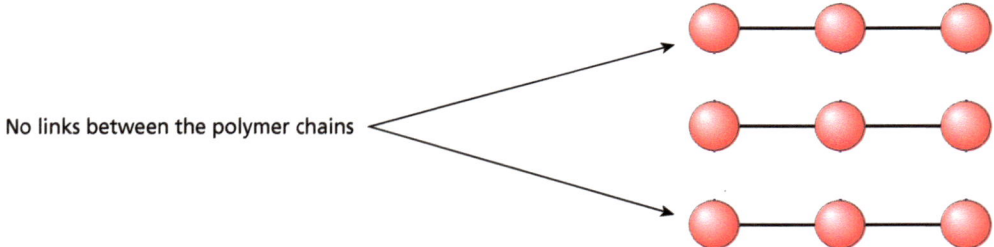

▲ Thermoplastics can be reshaped as there are no cross-links between the polymer chains, so the plastics become mouldable when heated.

The following section details some common examples of thermoplastics.

Acrylic
Also known as Perspex®, acrylic is quite hard wearing and can be used for many applications such as signs, glass-substitute phone covers and kitchenware.

High-impact polystyrene (HIPS)
This is a tough, rigid plastic that is often used in the vacuum-forming process. It is good for packing (biscuit tin inserts), toys and cutlery dividers/drawer organisers.

▲ Nylon can be used for cogs because of its low friction properties.

PVC
PVC is a hard-wearing plastic used for doors and windows (as **UPVC**), waste pipes, guttering and electrical tape. It is also used for many other products such as plumbing fittings, electrical wiring and products used in the medical industry.

Nylon
Nylon has excellent low-friction properties. It is great for assemblies with moving parts, especially for products such as door runners, cogs/gears and washers. Nylon fibres are incredibly strong and can be wound into ropes.

▲ As polypropene has a high melting point, it can be used to make kitchen utensils.

Polyesters
These are very strong and hard wearing and they are used in textiles for clothing or cleaning cloths. Polyester fabric is also used as reinforcement for rubber in drive belts, rubber tyres and hoses.

Polypropene
This has high strength, good flexibility and a high melting point. It can be moulded into kitchen utensils, tubes and pipes. It can also be produced as a fibre for making ropes.

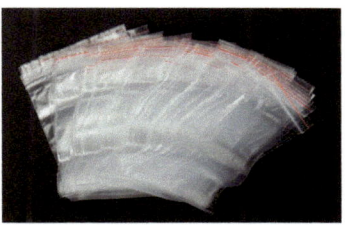
▲ Polythene is commonly used for bags or for wrapping and packaging other products.

Polyethene (polythene)
Polyethene (or polythene) is tough and flexible. It is commonly used as a thin film for wrapping and packaging, but it can also be used in containers, pipes and mouldings.

Polyphenylethene (polystyrene)
Known more commonly as polystyrene, polyphenylethene can be used as foam packaging and for disposable drinking cups. It can also be made into hard, tough and rigid mouldings such as those used in refrigerator interiors.

▲ Expanded polystyrene is used to make cups for hot drinks as it is a good thermal insulator.

Polytetrofluoroethene (Teflon)
Teflon is a tough, heat-resistant material. It has a very smooth surface with low coefficient of friction making it ideal as a bearing material. Teflon is commonly used for the non-stick coating on cooking utensils and in sheet or tape form for seals and gaskets.

▲ Teflon is used to make the non-stick surfaces of cooking pans.

> **Key Terms**
>
> **PVC** A type of thermoplastic. Its full name is polyvinyl chloride.
>
> **UPVC** Unplasticised polyvinyl chloride. This hard form of PVC is often used for doors and windows.

Task 8.4

1 Explain the term 'thermoplastic', with reference to molecular structure.
2 Thermoplastics are very useful in the manufacture of mobile phone casings. Research the name of a thermoplastic used for this purpose.
3 Describe the properties of the polymer you have named above that make it suitable for mobile phone casings.
4 In your notebooks, match the name of the thermoplastic to its use.

Polystyrene   PVC   Polypropene   Polyester   Teflon   Polythene   Nylon

Thermosetting plastics

Unlike thermoplastics, thermosetting plastics have cross-links between the polymer chains. This gives thermosetting plastics a very strong bond between the monomers. Once 'set', these plastics cannot be reheated to soften, shape and mould them, as thermoplastics can. As a result of the molecules of these plastics being cross-linked in three dimensions, they cannot be reshaped or recycled.

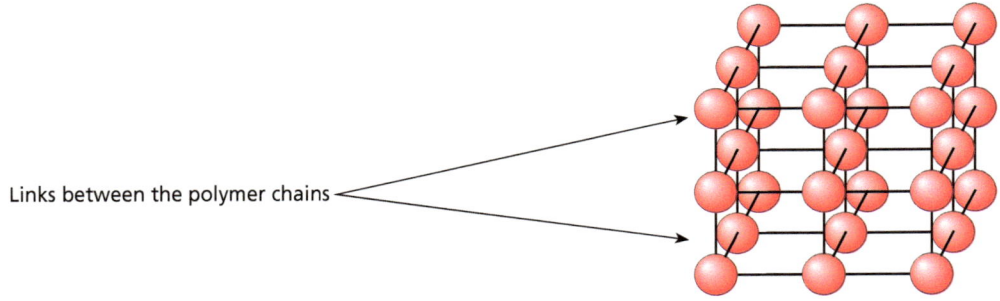

Links between the polymer chains

▲ Thermosetting plastics cannot be reheated and reshaped as a result of the strong cross-links between the polymer chains.

The following section details some common examples of thermosetting plastics.

Epoxy resin

Often used as a type of glue, epoxy resin is good for laminating (layering) materials to create products such as skateboards. It is also known by the brand name Araldite®.

Urea formaldehyde

This is a hard, slightly brittle plastic used for electrical casing/housing such as plug sockets and smoke alarms.

Melamine formaldehyde

Melamine formaldehyde is used for tableware, electrical insulation, laminates for worktops and synthetic resin paints. This is because it is rigid, heat resistant, scratch resistant, it has good strength and hardness, and can be coloured.

▲ Melamine is the plastic used for colourful plastic plates.

8 Understanding properties of engineering materials

Phenol formaldehyde
This plastic can be used for cheap electrical fittings, parts for domestic appliances, bottle tops and kettle/iron/saucepan handles. It has high heat resistance, but is very dark, hard and quite brittle.

Phenolic resin (Bakelite)
Bakelite is hard, heat resistant and solvent resistant. It is a good electrical insulator and can be machined but it has limits for decorative purposes. It is used for electrical components and also for heat-resistant handles such as those on kettles and saucepans.

Polyester resins
These have similar properties to epoxy resins, with good heat resistance and a hard-wearing surface. They are moulded with the same glass fibre, carbon fibre and Kevlar reinforcement materials. Boat hulls, skis, motor panels, aircraft parts and fishing rods can be moulded from these reinforced resins.

Urea-methanal resin (Formica)
Formica is naturally transparent, but can be produced in a range of colours. Kitchen worktops and worktable tops are often covered with a layer of hard-working Formica laminate. It is also used for items of kitchenware, toilet seats and electrical fittings.

▲ *Formica is often used to make kitchen worktops.*

Task 8.5

1. Explain the term 'thermosetting polymer', with reference to molecular structure.
2. Thermosetting plastics are very useful in the manufacture of electrical fittings. Name a thermosetting polymer used for this purpose.
3. Describe the properties of the polymer you have named above that make it suitable for electrical fittings.
4. In your notebooks, match the name of the thermosetting plastic to its use.

Polyester resins Phenol formaldehyde Urea formaldehyde Bakelite Formica Epoxy resins

Composite materials

Composite materials, or composites, are made from two or more materials. They are combined in order to add extra properties such as increased strength. Unlike alloys, composite materials are not combined after melting or mixing but are kept separate and are usually bonded together with glues, adhesives or resins. The different materials work together to produce a new material which combines all of the properties of the previously separate materials. Within the composite it is still possible to tell the different materials apart easily; they do not tend to blend or dissolve into each other. Composite materials can be either man-made or they may exist in nature.

Below is a  of a composite front door. Many homes now have these types of composite products/materials as their exterior doors and the annotations explain why.

> **Key Terms**
>
> **Composite** Something made up from several parts or elements.
>
> **Cutaway drawing** A drawing designed to show the viewer important parts of the interior of an opaque object or product.

WJEC Level 1/2 Vocational Award Engineering (Technical Award)

> **Key Terms**
>
> **GRP** Glass-reinforced plastic, also called fibreglass. A mix of fibreglass and epoxy resin.
>
> **MDF** MDF, or medium-density fibreboard, is a man-made board used for many applications such as furniture and internal building projects.

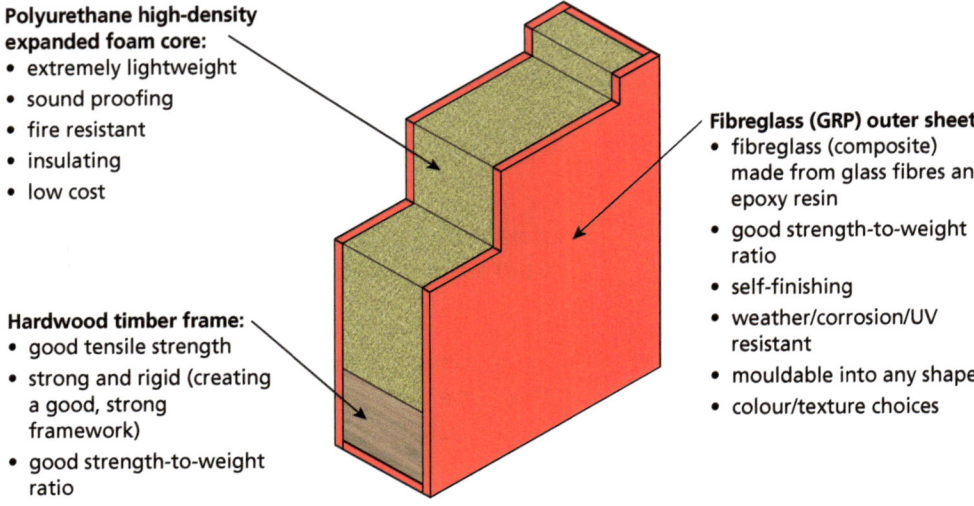

Polyurethane high-density expanded foam core:
- extremely lightweight
- sound proofing
- fire resistant
- insulating
- low cost

Hardwood timber frame:
- good tensile strength
- strong and rigid (creating a good, strong framework)
- good strength-to-weight ratio

Fibreglass (GRP) outer sheet:
- fibreglass (composite) made from glass fibres and epoxy resin
- good strength-to-weight ratio
- self-finishing
- weather/corrosion/UV resistant
- mouldable into any shape
- colour/texture choices

▲ A composite front door

There are many types of composite material that are commonly used in everyday products. Examples of composite materials include:
- man-made/manufactured boards
- GRP
- carbon fibre
- reinforced concrete
- Kevlar.

Man-made/manufactured boards

Made from timber off-cuts, chips, fibres and glues/adhesives, manufactured boards include plywood, chipboard, MDF and blockboard. They are strong, resistant to warping, low cost and can come in large-surface-area sheets/boards. They are commonly used for furniture in stores such as IKEA.

Chipboard is made by gluing together wood particles with an adhesive under heat and pressure. This creates a rigid board with a relatively smooth surface. Chipboard is available in a number of densities: normal, medium and high density. It is a very common material used in building works due to its properties and cost.

GRP

GRP is made from glass fibres and epoxy resin and is used in many products such as boat hulls, waterslides, wind turbine blades and telecoms street cabinets (big green units). It has a good strength-to-weight ratio, can be moulded to almost any shape, is relatively lightweight and is UV resistant. Each individual glass fibre is very fine with a small diameter. The strands or fibres are woven to form a flexible fabric. The fabric can then be placed in a mould, for instance a mould for a canoe, and polyester resin is added, followed by a catalyst (to speed up the reaction).

Carbon fibre

Products made from carbon fibres and resins are used in sports equipment, motorsports (F1), the aerospace industry and safety equipment. The material is extremely lightweight, has a fantastic strength-to-weight ratio and can be moulded to almost any shape.

Reinforced concrete

On its own, concrete is weak when put under tension. If reinforced with steel, however, it is stronger and can withstand tensile forces. Concrete's relatively low tensile strength and ductility are counteracted by the inclusion of the steel which is strong and ductile.

▲ Man-made boards

▲ You can see the wood particles making up this section of chipboard.

▲ Carbon fibre

Kevlar

Kevlar is a liquid converted into a fibre and woven into textile materials. It is strong, lightweight, corrosion resistant and heat resistant. Kevlar has a high tensile strength to weight ratio. It is commonly used for bulletproof jackets, armour for military vehicles and planes, and F1 fuel tanks.

▲ *Kevlar is used in bullet-proof jackets.*

> ### Task 8.6
> 1. What is a composite material?
> 2. How are the physical properties of a composite material better than those of a single material?
> 3. Describe one advantage of the composite steel-reinforced concrete over normal concrete.
> 4. What composite material is used for the manufacture of canoes, boat hulls and waterslides?
> 5. Describe the composite material you have named above and explain how it is manufactured.
> 6. Describe the physical properties of Kevlar and list some uses of this composite material.
> 7. Find out what CFRP is. What properties does it have and where is it used?

Smart materials

Smart materials are materials that can change their properties and characteristics according to changes in the environment in which they function. This means that one of their properties can be changed by an external condition, such as temperature, light, pressure or electricity. This change is reversible and can be repeated many times. Smart materials are being used more often in products and processes as they can complete multiple tasks due to their changing nature.

Examples of smart materials include:
- nitinol
- D3O®
- polymorph
- QTC.

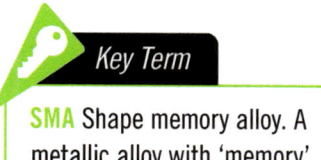

Key Term

SMA Shape memory alloy. A metallic alloy with 'memory'.

Nitinol
Nitinol is an **SMA** that can 'remember' a shape when heated. It is an alloy of nickel and titanium and is used for stents (small devices used to hold open arteries), underwire bras and flexible frames for glasses.

D3O®
This is a lightweight, soft, flexible and malleable material that stiffens when subjected to sudden force. It is commonly used for sports clothing, including skaters' beanies (headwear) and motorbike riders' 'leathers'.

Polymorph
Polymorph is a hard polymer (plastic) which comes in the form of granules but which softens and becomes malleable at 62°C. It can be used for tool handles and individual grips.

QTC (quantum-tunnelling composite)
QTC is a flexible polymer which contains tiny metal particles. It is normally an insulator but if it is squeezed it becomes a conductor. QTC is often used for switches on mobile phones, pressure sensors and speed controllers.

▲ *A biker's jacket with D3O® panels*

▲ *Polymorph granules*

Key Term

Photochromic A substance that changes colour according to the light; *photo* = relating to light, *chromic* = relating to colour.

Thermochromic A substance that changes colour according to the temperature; *thermo* = relating to temperature, *chromic* = relating to colour.

Hydrochromic A substance that changes colour according to the level of water; *hydro* = relating to water, *chromic* = relating to colour.

Pigment A coloured molecule.

Aroma Scent or smell.

The following sub-sections list some general categories of smart materials.

Photochromic materials

Photochromic materials, like photochromic inks, darken as the light level increases. Some of these inks can even change colour. These smart materials react to changes in light levels and can be found in glasses that darken in bright environments. They are also used for security markers that can only be seen in ultraviolet light.

▲ Glasses containing photochromic materials go dark as the light level increases.

Thermochromic materials

Have your ever seen a mug change colour when hot water is poured in?

▲ This colour-changing mug is possible because of thermochromic materials.

The environment of the mug has been changed by the addition of heat. The so-called **thermochromic** ink used on the mug changes colour according to the temperature.

Thermochromic materials are also commonly used for contact thermometers made from plastic strips and food packaging materials that show you when the product they contain is cooked to the right temperature.

Hydrochromic materials

Hydrochromic inks change colour when they come into contact with water. A plastic moisture tester can be pushed into the soil alongside a plant. If the water content of the soil is at the right level, the colour of the moisture tester should remain blue. However, if the soil becomes dry, the colour changes to yellow.

Aroma pigments

These **pigments** (inks or paints) produce an **aroma** when scratched. They are popular in 'scratch-and-sniff' products, such as the perfume samples often found in magazines.

Piezoelectric materials

When a **piezoelectric** material is squeezed rapidly, it produces a small electrical voltage for a moment. If a voltage is put across the material, it results in a tiny change in shape. Piezoelectric materials are often used in contact sensors for alarm systems, microphones and headphones.

▲ Headphones contain piezoelectric materials.

8 Understanding properties of engineering materials

Electroluminescent materials
Electroluminescent materials give out light when an electric current is applied to them. They are used for safety signs and clothing for use at night.

Magneto-rheostatic (MR) fluids and electro-rheostatic (ER) fluids
Both of these materials can change rapidly from a thick fluid to an almost solid substance and back again in milliseconds. MR fluids change when exposed to a magnetic field, while ER fluids change if exposed to an electric field. The fluids contain microscopic metal particles suspended in a type of oil. In the environment of a magnetic or electric field, the particles immediately line up which greatly restricts the movement of the fluid. They can be used for fast-acting clutches, shock absorbers and flow-control systems.

▲ *Magneto-rheostatic fluids*

Hydrocarbon-encapsulating polymers (HC polymers)
These are polymers that absorb oil, forming a rubbery substance. They are environmentally 'friendly' and have been developed to manage hydrocarbon-based liquid spills. A potential practical application is at petrol/diesel filling stations. If there is a spill at the pump, which could be hazardous, an HC polymer can be applied and this absorbs the fuel safely. It can then be used as a solid fuel and even burned.

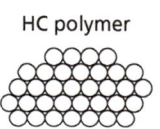
HC polymer

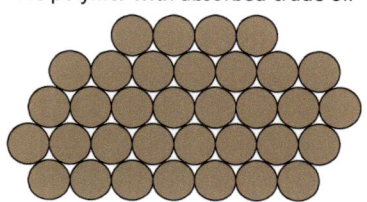
HC polymer with absorbed crude oil

▲ *HC polymers can be used to clean up oil spills.*

Nanomaterials
Nanoparticles are tiny particles (generally less than 100 nm in diameter) that can be used to improve the mechanical properties of a material, such as stiffness or elasticity. They are often used in car manufacturing to create cars that are faster, safer and more fuel efficient. Nanomaterials can also be used to produce more efficient insulation and lighting systems.

Task 8.7

1. In your notebook, copy and complete the table below naming the smart material used for each application.

Wireless headphones	
Food packaging that shows you when a product is cooked to the right temperature	
A 'scratch and sniff' perfume sample in a magazine	
Reflective stop sign	
Oil spill clean-up	
Security marker pens for bank notes	

2. What is the definition of a smart material?
3. What is polymorph and how can it be used?
4. What does SMA stand for?
5. What are thermochromic inks? Include a description of one practical application.
6. Name another smart material and describe a product in which it is used.
7. Describe the following smart materials:
   a nanomaterials
   b aroma pigments
   c magneto-rheostatic fluids.

Task 8.8

Copy the following tables into your notebook and use the knowledge you have gained about materials and their properties in this chapter to complete them.

Ferrous metals		
Name	Product it is used for	Properties needed

Non-ferrous metals		
Name	Product it is used for	Properties needed

Alloys			
Name	Parent metal	Product it is used for	Properties needed

Polymers/plastics			
Name	Thermoplastic or thermosetting	Product it is used for	Properties needed

Smart materials		
Name	Product it is used for	Properties needed

Composite materials		
Name	Product it is used for	Properties needed

Describe properties required of materials for engineering products

Materials and how they perform are defined by their properties. Material properties tell an engineer what that material is good at and what it is not so good at. For an engineer to provide a competent solution to the design brief, they must have an understanding of how the materials perform in different conditions to ensure the right material is selected for the job.

Common engineering properties

The following section describes some common engineering properties.

Tensile strength

Tensile strength is a measurement of the force required to pull something, such as a rope, wire or a structural beam, to the point where it breaks. The cables holding up a suspension bridge are made of steel because they bear the weight of the road and the traffic and can retain their strength when being stretched. We would say that steel has high tensile strength.

Compressive strength

This is a measure of how well a material maintains its strength when under a heavy load or pressure. Large office buildings will have concrete under stress, but will be reinforced to ensure they do not collapse under this stress.

Hardness

Hardness is a measure of how resistant a material is to various kinds of permanent shape change when a compressive force is applied. It is the ability of a material to resist scratching, cutting or wear and tear. Some materials, such as metals, are harder than others.

Toughness

Toughness is the ability of a material to absorb energy and deform without fracturing. It can also be defined as the amount of energy that a material can absorb before it ruptures or breaks, or a material's resistance to fracture when stressed. The toughness of mild steel means that it can be used to absorb the impact of a crash in a car crumple zone, a safety feature of all modern cars.

Malleability

Malleability is the ability of a material to deform under pressure (compressive stress). A malleable material may be flattened by hammering or rolling. Gold has a good malleability as is shaped and rolled into decorative jewellery and ornaments with relative ease.

Ductility

Ductility is when a solid material stretches under tensile stress. A ductile material may be drawn or stretched into a wire. Most metals are good examples of ductile materials, including gold, silver and copper.

Electrical conductivity

Electrical conductivity defines a material's ability to conduct electricity. Electric current can flow easily through a material with high conductivity. Metals have high conductivity, while rubber has a very low conductivity. Factors such as temperature have a large effect on conductivity.

Thermal conductivity

The thermal conductivity of a material is a measure of how quickly energy transfers through it by heating. A material with high thermal conductivity transfers energy quickly, while a material with low thermal conductivity transfers energy slowly, and may be used as a thermal insulator.

Corrosion resistance

Corrosion resistance refers to how well a substance (especially a metal) can withstand damage caused by oxidisation or other chemical reactions. An example of corrosion resistance is when a boat is treated to prevent rust and is thus able to withstand damage.

▲ *The cables holding up this bridge need very high tensile strength.*

Key Term

Strength The capacity of an object or substance to withstand great force or pressure.

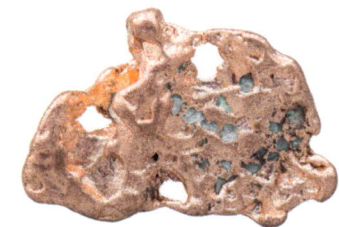

▲ *A ductile material like copper metal (top) can be drawn into wires (bottom).*

▲ Mountain bike tyres need good elasticity.

Environmental degradation

If a material can resist environmental degradation, it may be able to be used in specific settings. For example, many rotary washing lines are made from aluminium. Not only is aluminium light but it does not corrode or rust (there is no iron in it) and it resists degradation from UV rays (the Sun). It is therefore a great material to use outdoors.

Elasticity

Elasticity is the property of solid materials to return to their original shape and size after the forces being applied to them have been removed. Mountain bike tyres have excellent elasticity as they need to deform and change shape when going over rough terrain. They also change shape when flat, but return to the original shape when pumped up.

Task 8.9

When creating engineered products, it is important for a designer to understand what materials are required and why. Will steel rust if used for a bridge? Will aluminium hold a human weight for a stool? Knowing and understanding the properties of engineering materials is essential when designing any product.

Below are four images of four very different engineered items comprising materials chosen for their specific properties.

Sketch each of the engineered items into your notebook on a separate page. Add annotations around them to explain what properties the various parts need to have to be successful. Evaluate your answers and explain your reasoning.

At the bottom of each page, try to rank the three most important properties you have identified into gold, silver and bronze, which gold being the most important. Explain why you have put the properties into that order. You can arrange these in three separate boxes or just use separate paragraphs or bullet points.

▲ A modern smart phone

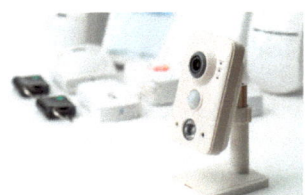
▲ A home security alarm

▲ A bicycle

▲ A children's play area

Task 8.10

Copy the table below into your notebook. Complete the table by naming a suitable material and product for the properties listed.

Property	Material	Product where it is used
Tensile strength		
Malleability		
Electrical conductivity		
Elasticity		
Hardness		

Explaining how materials are tested for properties

Testing of materials and their properties

When the process of selecting materials is complete, engineers begin the testing process to see whether the properties of the materials are sufficient to complete the task asked of them. Sometimes the completed solution/product is tested in its functioning environment.

It is very important that engineers test materials for the correct properties, as selecting the wrong materials or inadequate properties could result in costly mistakes. Imagine a bridge collapsing because the materials were not tested properly.

Materials are also tested to define the limits of their operating parameters. In other words, how much load something can take before it starts to break. Once those parameters are understood, engineers can start designing into their product a **factor of safety (FoS)**. For example, when the failure point of a product/material is found, engineers can add more material to take double the load, essentially increasing the FoS. Testing the limits of a material and product allows engineers to ensure their engineered solutions are very safe to use.

By going through the testing process engineers can:
- save money
- meet health and safety needs/standards
- prevent product failures
- provide data for future projects/innovations.

Testing for materials' properties can fall under two categories: **destructive tests** and **non-destructive tests**.

Destructive tests

In destructive tests, materials (or products) are put under **force** until they begin to fail and then **catastrophically fail**. **Destructive testing (DT)** includes methods in which the material is broken down in order to determine its mechanical properties such as tensile strength, compressive strength, toughness and hardness. DT is a way of finding out the exact point at which a material or a product will fail. It means, for example, finding out whether the quality of a weld is good enough to withstand extreme pressure.

Benefits of destructive testing
- It verifies the properties of a material.
- It determines the quality of materials.
- It helps to reduce failures, accidents and costs.
- It ensures compliance with regulations.

Types of destructive testing
- Tensile strength testing
- Bend testing
- Impact testing
- Nick break testing
- Hardness testing

Some of these tests are discussed in greater detail in a later section in this chapter.

> ### Key Terms
>
> **Factor of safety (FoS)** To build in a safety margin when designing products. The idea that something is designed in such a way that it is safer than it needs to be.
>
> **Force** A push or pull on an object causing it to change velocity (to accelerate) or change shape.
>
> **Catastrophically fail** A product is tested until it is broken and does not work anymore.
>
> **Destructive testing (DT)** Testing materials until they break. Discovering the exact point at which a material or product will fail under testing.

▲ Destructive testing: testing a material or product until it fails.

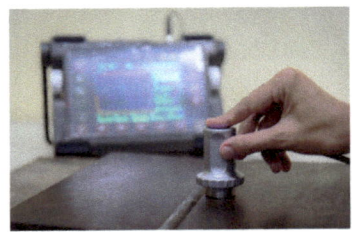

▲ Non-destructive testing

> **Key Term**
>
> **Non-destructive testing (NDT)** A non-intrusive way of testing materials.

Non-destructive tests

In non-destructive tests, the materials (or products) are tested without them becoming damaged. The process allows you to test your material/product in its functioning environment to see how it performs in its day-to-day life. Non-destructive testing can save money by not having to prepare and destroy materials and products. This testing process also allows for engineers to test the integrity of historical structures. **Non-destructive testing (NDT)** involves inspecting, testing or evaluating materials, components or assemblies for discontinuities, or differences in characteristics. You do not destroy the serviceability of the part or system. In other words, when the inspection or test is completed, the part can still be used.

Benefits of non-destructive testing
- It is cost effective as materials are not damaged.
- You can do more than one test on a product.
- It doesn't break materials.
- It provides real-time and accurate data.
- It is possible to inspect a wide range of materials.

Types of non-destructive testing
- **Visual inspection:** inspectors carefully examine the surface, looking for visible signs of damage such as cracks, corrosion or discontinuities. The process may involve the use of borescopes or other optical aids. While it is valuable for identifying surface-level issues, it may not detect internal defects or subsurface anomalies, so should be complemented by other NDT methods.
- **Magnetic particle inspection:** the material to be tested is magnetised using a strong magnetic field, and iron particles are applied to the magnetised surface. Defects attract particles, forming visible indications. This method is limited to ferromagnetic materials and only detects defects perpendicular to the magnetic field.
- **Dye penetrant testing:** a coloured dye (penetrant) is applied to the surface, penetrating into any surface defects.
- **Radiography:** X-rays or gamma rays pass through the object and a radiographic detector captures the images, revealing internal features and defects.
- **Ultrasonic flaw detection:** ultrasonic waves are sent through the material, and reflected waves are analysed for indications of defects.
- **Eddy current testing:** a coil induces electrical currents in the material. Changes in the current indicate defects.

Testing engineering properties

The following section provides brief details about some of the ways in which the properties of materials can be tested.

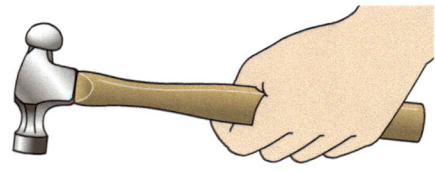

Hardness testing
A basic test of hardness involves using a centre punch to indent the surface of a material. Different materials require a different amount of force to form an indent. A big dent indicates a weak material; a small dent indicates a hard material.

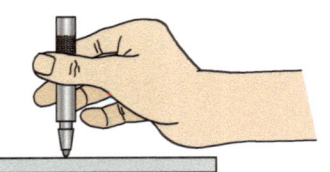

▲ Hardness testing using a centre punch

Tensile strength testing

To test the tensile strength of a material, it is clamped in a vice with a fixed weight hung from the end. The amount of deflection of the material under these conditions is a measure of its resistance and tensile strength. A selection of materials of the same section, cut to exactly the same size, can be tested in this way and so compared.

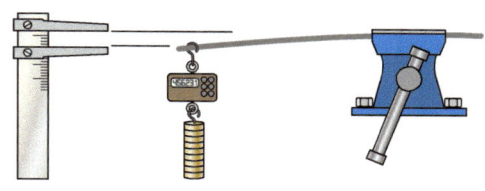
▲ *Tensile strength testing*

Toughness testing

A simple school workshop test for toughness involves hitting a sample of material with a hammer while it is secured in an engineer's vice. If it survives the blow, without bending too far, it can be said to be tough. If it shatters, it can be said to be brittle.

Malleability and ductility testing

A length of tube is placed over a piece of material and used as a lever. The material is then folded to an angle of 90°. Any cracks or damage on the inside of the bend represent a lack of malleability, while any cracks or damage on the outside of the bend represent a lack of ductility.

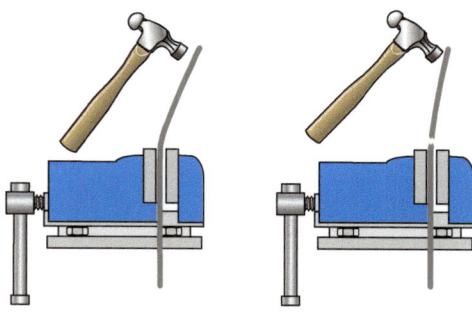
▲ *Toughness testing*

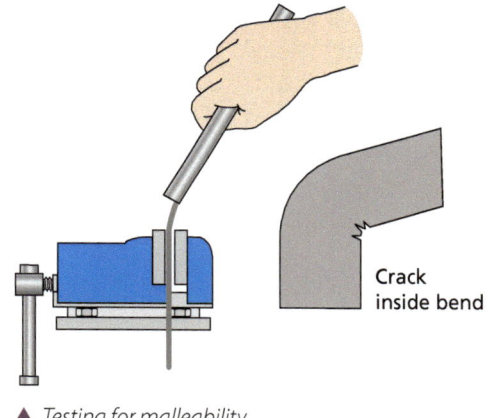

▲ *Testing for malleability*   ▲ *Testing for ductility*

Electrical conductivity testing

A voltmeter is used to measure resistance. The probes are set to the same distance on each sample of material and the resistance is measured. The value of resistance is a measure of the material's conductivity.

Thermal conductivity testing

A fan of rods made of different materials can be heated at one end with the same Bunsen flame. Whichever rod gets hottest at the other end first is the best conductor. This can be tested by having a small item like a pea or a metal tack attached to the far end of each rod with Vaseline. When the heat reaches the end of the rod, it will melt the Vaseline and make the item drop. The material that heats the quickest is said to have the highest thermal conductivity.

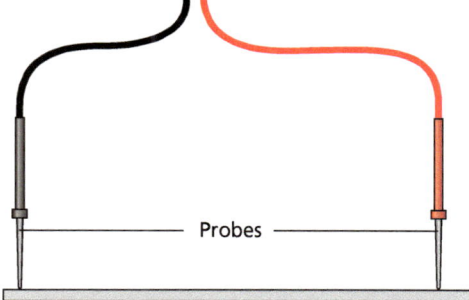
▲ *Testing the electrical conductivity of a material*

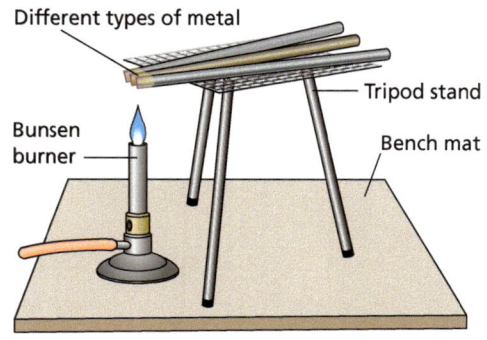
▲ *Comparing the thermal conductivity of three different metals*

Elasticity testing

A piece of material is clamped in a vice. A fixed weight is hung from the end and the material bends. When the weight is quickly removed, the material should 'spring back'. The amount of deflection is a measure of the elasticity of the material.

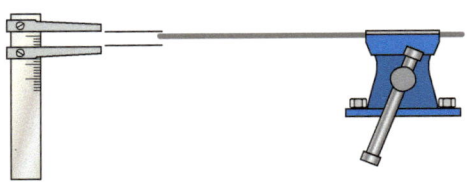

▲ *Elasticity testing*

Corrosion resistance testing

In corrosion testing, a material is subjected to simulated conditions and then carefully analysed for signs of cracking, fatigue, pitting, rusting and other damage.

CAD and testing

Most good computer-aided design programs now come with the facility to test materials and products in virtual environments. Modern CAD programs already contain data for materials. When designing your product, you will specify the materials your CAD models are made from. If the software deems that your product is going to fail under a load then it can show in colours (mainly red) exactly where your product would fail. This allows engineers to redesign their solutions before going into production. This process is called **finite element analysis (FEA)**.

Key Term

Finite element analysis (FEA) A way to test your material (element) of choice under forces on a computer model.

Task 8.11

You have been asked to measure the ability of a tennis racquet to resist bending. Describe, using notes and sketches, how you could perform a simple test in a workshop to measure how well the tennis racquet can resist bending.

Advice
- Break your test down into stages – 1, 2, 3, 4 etc. For example, Stage 1: place racquet in vice.
- Include quick diagrams with annotations.
- Make a list of equipment/tools required.

Task 8.12

Using notes and sketches, describe how the hardness of a mild steel bar could be tested.

Advice
- Break your test down into stages – 1, 2, 3, 4 etc. For example, Stage 1: place metal in vice.
- Include quick diagrams with annotations.
- Make a list of equipment/tools required.

Guide to coursework submission: Unit 2

9

This chapter look at the second unit that you need to complete, and examples of tasks and pages that you can complete ready for submission.

Unit weighting

The WJEC Level 1/2 Vocational Award in Engineering is broken up into three different units. In each unit you will be asked to demonstrate your engineering skills and knowledge, and create evidence via portfolios, manufacturing skills and examinations. The evidence then needs to be submitted in an electronic format (your tutors will complete this part).

Each unit will have a different task and a different weighting:

Unit number	Evidence required	Weighting of unit	Time to complete (per student)
Unit 1	Manufacture a prototype Produce an electronic portfolio	40% (80 marks)	20 hours
Unit 2	Design/redesign part of prototype Produce an electronic portfolio	20% (40 marks)	10 hours
Unit 3	Complete a controlled examination	40% (80 marks)	1 hour 30 minutes

Examples for submission

Some of the difficulty in producing work for a technical subject such as engineering lies in understanding what you need to submit as evidence. The guidance offered by exam boards, although detailed, needs to be translated into actual tasks you can do.

The following examples are tasks published by WJEC that you could complete and that would satisfy the needs of the assessment grids (marking structure) for Unit 2.

IMPORTANT: The following are **EXAMPLES ONLY**. You must translate the needs and requirements of each unit with **YOUR OWN WORK**. You can use these examples as **GUIDANCE ONLY** and should not submit copies of the following work if you want to ensure you are not questioned about the originality of your submissions.

Unit 2: submission guide

In this guide you will find examples of evidence on what to create and submit for Unit 2 Designing engineering products.

This guide is made up from examples. You should interpret the needs of the specification yourself and produce your own work.

Brief example

Assignment brief

Following on from the success of the LED lamp, the lighting company has received feedback from customers with suggested improvements to the design. Some of the more common suggestions are: reducing the overall weight of the lamp and looking at ways in which it can be clamped down to a tabletop.

You have been asked to redesign all or part of the lamp to fit the requirements of the following specification. You can use the information you developed in Unit 1 to help with the process.

This is a design task only with no need to manufacture new parts.

Design specification

- The modified solution must hold the lamp securely to a desk or table.
- The design should continue to allow the lamp to be rotated to allow light to be shone at any angle.
- The clamp solution must not harm the desk or table (e.g. no drilling of holes into the desk).
- The lamp could include new dimensions for some parts to reduce the weight/material used.
- The clamp mechanism must be hand-operated.

Explain the individual functions of the primary features of the product.

You should consider (where appropriate):
- electrical components
- mechanical components
- component properties.

[2 marks]

Example page

Identify the different parts of the product you manufactured in Unit 1 and discuss the various functions of each part along with the materials it has been manufactured from and their properties.

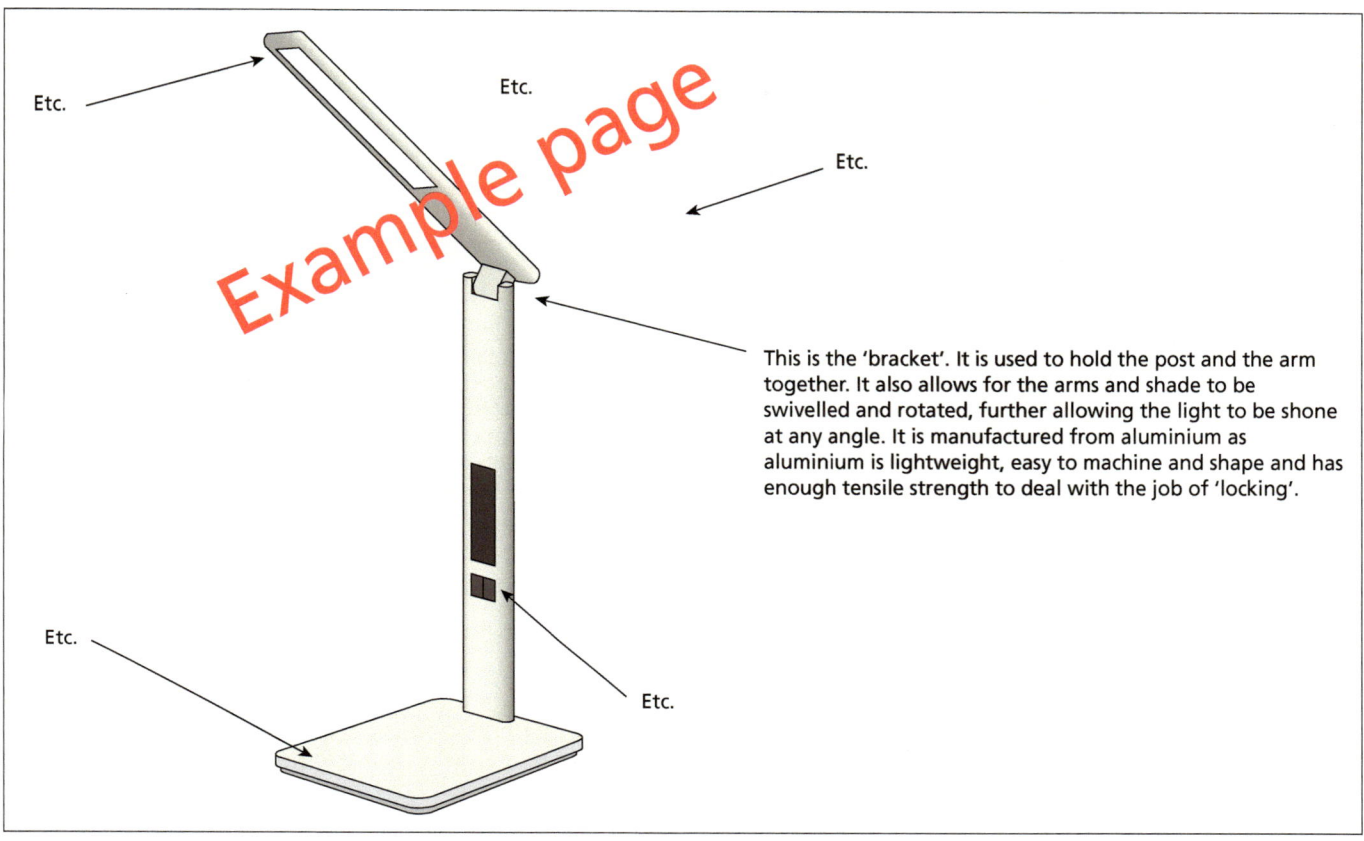

Etc.

Etc.

Etc.

This is the 'bracket'. It is used to hold the post and the arm together. It also allows for the arms and shade to be swivelled and rotated, further allowing the light to be shone at any angle. It is manufactured from aluminium as aluminium is lightweight, easy to machine and shape and has enough tensile strength to deal with the job of 'locking'.

Etc.

Etc.

Suggest at least two other engineered products that have similar functional properties to those required by the given brief.

[2 marks]

Justify how the functional properties of the found engineered products meet the requirements of the brief.

[5 marks]

Example page

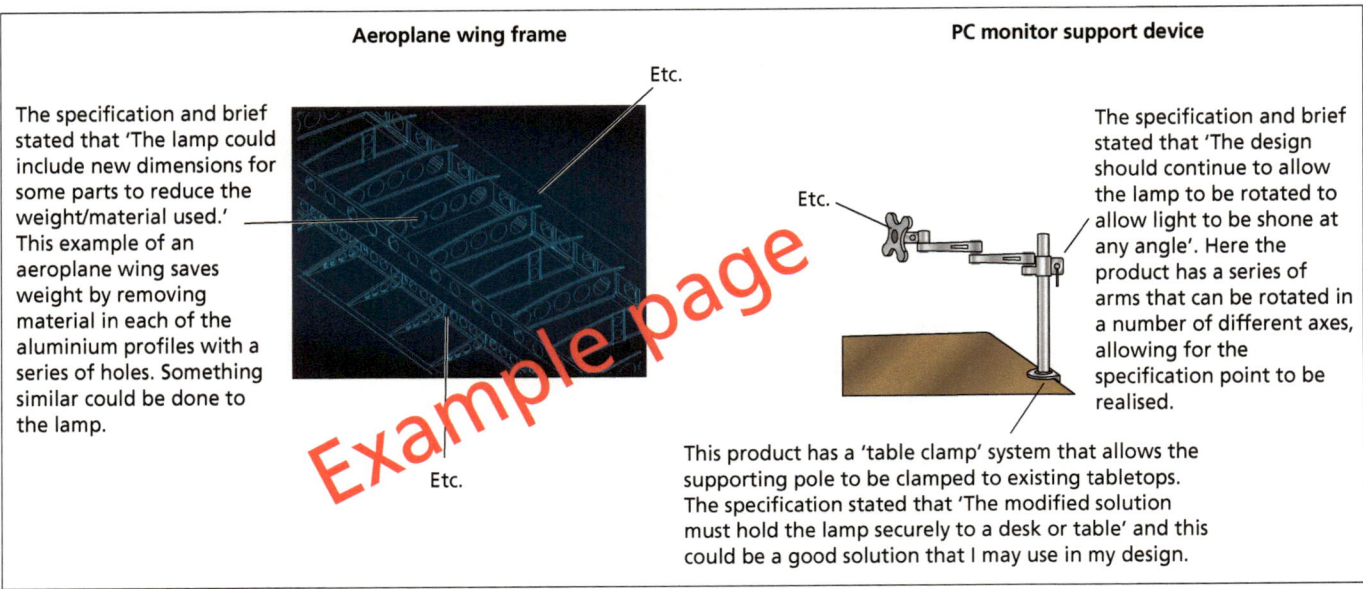

▲ Examples of two other engineered products which may have design features that could be applied to the lamp.

Make sure your annotations relate to **all** specification points for **both** products.

Design a range (3) of solutions that meet the brief and design specification.
This should include:
- identified features that meet the brief
- use of models (also CAD models) to support, develop and test the functional qualities of your ideas.

[4 marks]

Present design ideas clearly using suitable media appropriate to the information being displayed.

You should consider:
- conveying meaning
- using appropriate language
- having a logical structure
- clearly presenting the information using either ICT or traditional hand-written/illustration methods
- using appropriate terminology
- including visual support such as simple models, CAD visuals or test rigs.

[4 marks]

Examples

If you want to produce models then you **must** take pictures of them **and** show how they could be used or tested.

Your annotations should **ALWAYS** mention the **assignment brief** and the **specification**.

You should evaluate how your designs meet the criteria set in the brief and design specification relating to:
- materials
- sizes
- tolerances
- cost
- operational parameters

and recommend the best solution.

[4 marks]

Example page

Design idea	Specification point	Why it does meet this point	Why it doesn't meet this point	Does it meet the design brief? Why?
Idea 1 or A	The modified solution must hold the lamp securely to a desk or table.	Idea 1 uses aluminium for the jaws. This is appropriate because the properties are … The post is made from mild steel. This is appropriate because …	Etc.	This idea partially meets the brief because …
	The design should continue to allow the lamp to be rotated to allow light to be shined at any angle.	Etc.	Etc.	
	The clamp solution must not harm the desk or table (e.g. no drilling of holes into the desk).	Etc.	Etc.	
	The lamp could include new dimensions for some parts to reduce the weight/material used.	The post is of a thinner diameter and is made from aluminium, therefore …	Etc.	
	The clamp mechanism must be hand-operated.	Etc.	Idea 1 does not meet this point because the overall cost is …	
Idea 2 or B	The modified solution must hold the lamp securely to a desk or table.	Etc.	Etc.	This mostly meets the brief because …
	The design should continue to allow the lamp to be rotated to allow light to be shined at any angle.	Etc.	Etc.	
	The clamp solution must not harm the desk or table (e.g. no drilling of holes into the desk).	Etc.	Etc.	
	The lamp could include new dimensions for some parts to reduce the weight/material used.	Etc.	Etc.	
	The clamp mechanism must be hand-operated.	Etc.	Etc.	

Best solution: Idea 3

Reason: The best solution that meets the brief is Idea 3 because …

Develop chosen idea and use annotations that discuss the needs of the:
- specification
- brief.

Include visual evidence such as sketches (isometric), simple models, CAD visuals or test rigs.

Your annotations should **always** mention the **assignment brief** and the **specification**.

Create a manufacturing specification.

Outline an engineering specification that addresses key points required to produce the design solution.

[3 marks]

▲ *CAD visual*

Example page

Part	Material	Stock-form/sizes	Tool requirements	Machine/equipment requirements	Processes used	Safety	Tolerances	Finish
Bracket	Aluminium	9 mm × 3 mm flat bar	Steel rule, Engineer's square, Scriber, Centre punch, Hacksaw	Pillar drill	Drilling, Filing, Cutting		± 0.5 mm	Machine finished, Polished/buffed
Post	Aluminium	Etc.	Etc.	Etc.			Etc.	Etc.
Arm	Aluminium							

Try to include all the things that would be needed to manufacture the product. Imagine you gave this to a person on the street. Could they make it based on the information you have given?

You should draw, using conventions, engineering drawings of your final design solution. To do this, you must:
- include third-angle dimensioned orthographic views of the product
- include an isometric view.

[6 marks]

Example page

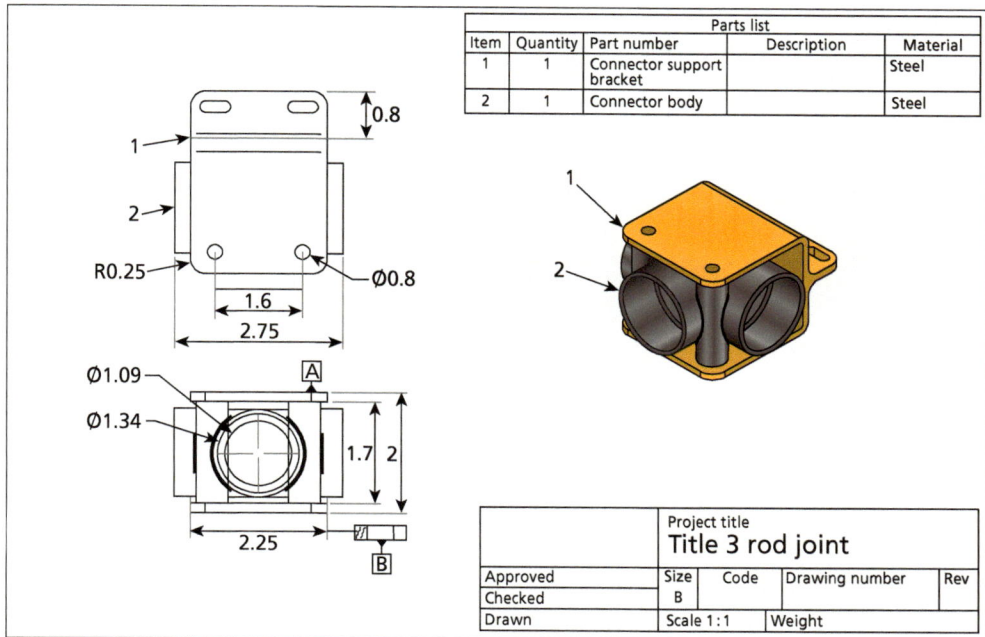

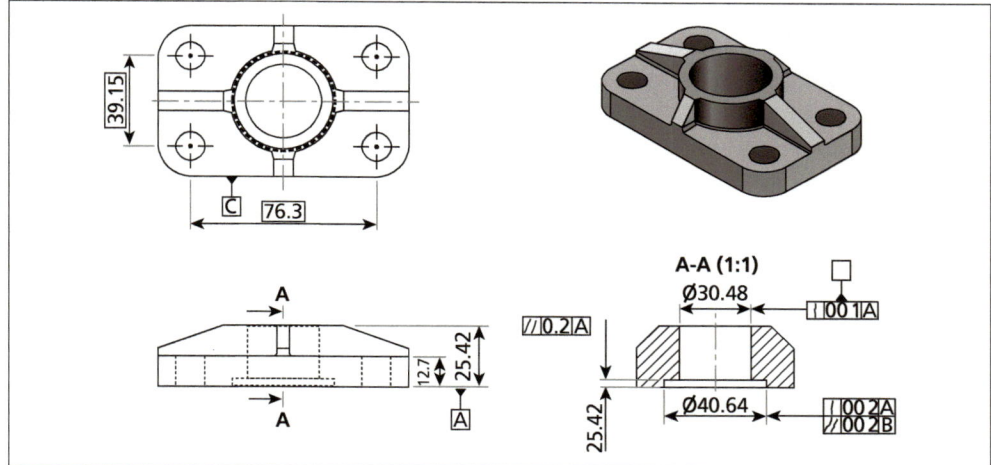

You **must** include:
- dimensions
- hidden detail lines
- section view
- isometric view
- table of parts and materials
- completed title block
- third angle symbol.

Example page

Final design
Use this page to show your final design in its **best light**.

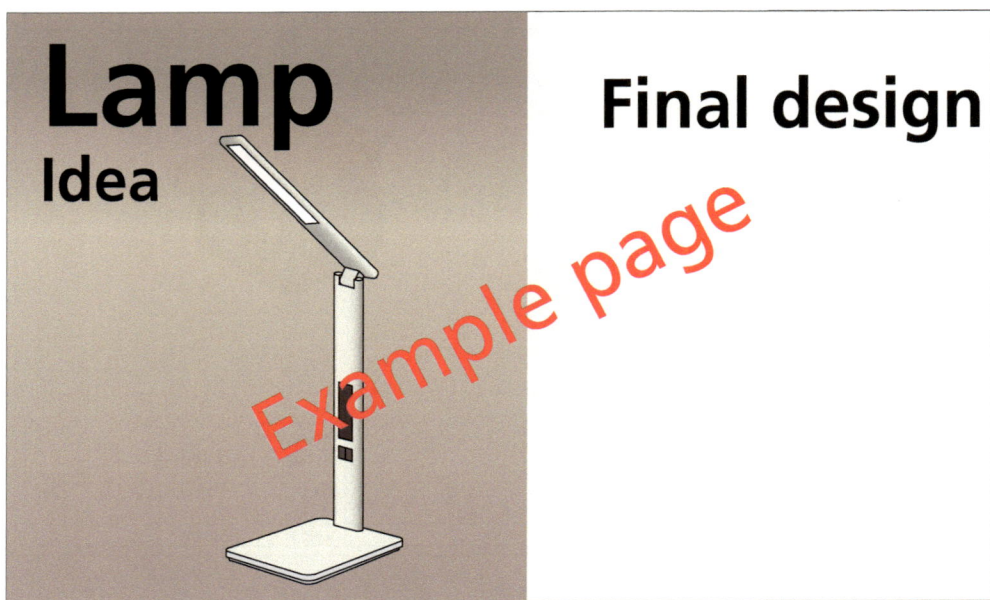

Apply mathematical techniques to determine specific problems identified in the given brief. To do this, you must:
- show all calculations
- use the correct units
- use mathematical conventions.

[4 marks]

Example page

Aluminium bracket		
Material	Size	Cost per bracket
Aluminium	45 mm (31.7 mm × 9.5 mm square section)	£0.72

Use this page to work out the total cost of the product. This **must** include all component parts.

This evidence may already be covered in previous work via:
- product analysis results
- design specification
- idea generation
- evaluations.

Ensure evidence is available to cover the requirements stated.

Advise the third party manufacturing the prototype about materials that may be used in the manufacture of the design modification. Justify your choices according to properties and testing results. Advise the manufacturer of the processes for manufacturing the component parts of the modified design, considering material removal and shaping, joining and assembly details, heat and chemical treatment methods, and finishing details where appropriate. Justify your choices.

[6 marks]

Glossary of key terms

Aesthetic The way something looks to the eye.

Aesthetics How a product or item may look. Is it aesthetically pleasing (nice to look at)?

Anthropometrics The science that defines physical measures of a person's size, form and functional capacities.

Aqueduct A bridge that carries water from one place to another.

Aroma Scent or smell.

Assembled Put together.

Axis The direction of travel from a fixed point. In 3D drawing there are three axes: X, Y and Z. (The plural of axis is axes.) A fixed reference line for the measurement of coordinates.

Axonometric perspective A pictorial representation of a 3D object that is not a true view of how you would view it.

Baseline The horizontal line you use to 'level' your set square.

Billet Also known as a billet of metal, a billet is a piece of metal of a certain size that is shaped by the forging process.

Bought-in component A component purchased from another manufacturing plant that does not need to be produced/manufactured.

The BSI Kitemark This mark is awarded by the BSI when a product meets its standards.

CAD (computer-aided design) Computer software for designing products in readiness for CAM.

CAM (computer-aided manufacturing) Machines manufacturing parts and products using a computer program.

Capillary action When liquid flows through very narrow spaces. In the brazing example, the **molten** metal is the liquid flowing through the space between two touching pieces of steel.

Catastrophically fail A product is tested until it is broken and does not work anymore.

Chart A way of displaying data or information. It can take the form of a table, a graph or a diagram.

Chasing The act of re-cutting a thread with a tap or die to repair any damage such as cross-threading. The process can also be used to clean up the thread if it is old, worn or dirty.

CNC Computer numerical control.

Composite Something made up from several parts or elements.

Compression A force that squashes or squeezes things together.

Condensed A condensed brief has been reduced in such a way that anything not required is taken out.

Conductivity A measure of the ease with which heat or electricity can pass through a material.

Constraint A limitation in a certain situation.

Construction lines Faint, thin lines that are easy to rub out and that can be used as a guide. They are drawn with a hard (H) pencil.

Conventions Technical terms.

Corrosion A gradual deterioration of metals caused by the action of air, moisture or a chemical reaction (such as an acid) on their surface. Corrosion can be described as the oxidisation of a metallic surface. When an iron surface corrodes, we say it has rusted.

COSHH Control of Substances Hazardous to Health. A set of regulations drawn up by the Health and Safety Executive (HSE) in 2002 to provide guidelines on the safe handling of hazardous substances.

Crate The name for the 3D 'box' you start your isometric drawings with.

Criteria Specific headings or titles.

Cutaway drawing A drawing designed to show the viewer important parts of the interior of an opaque object or product.

Datum A fixed starting point of a scale of operation.

Degradation The process of changing to a worse condition.

Destructive testing (DT) Testing materials until they break. Discovering the exact point at which a material or product will fail under testing.

Efficiency The state or quality of being efficient, or able to accomplish something with the least waste of time and effort. In scientific terms, efficiency is the relationship between the total energy input to a system and the useful energy output. The higher the useful energy output, the more efficient the system. Efficiency is calculated as a percentage.

Electrode An electrical conductor that is generally used to make contact with a non-metallic part of a circuit. In MIG and arc welding, the electrode is the sacrificial metal wire or rod.

Ellipse A circle viewed in axonometric projection.

Environmental Relating to or caused by the surroundings in which someone lives or something exists.

Ergonomics An applied science concerned with designing and arranging things people use so that the people and things interact efficiently and safely.

Expendable mould A temporary mould that will be destroyed when the casting process is complete. It is not reusable.

Extension lines These lines show the extent of the area that is being dimensioned. They are also known as lead lines.

Extrusions Profiles that have been extended or stretched.

Fabricate To manufacture something from different parts; to shape and join materials to create a product.

Factor of safety (FoS) To build in a safety margin when designing products. The idea that something is designed in such a way that it is safer than it needs to be.

Feasible Manageable or possible.
Ferrous A ferrous metal is any metal that contains iron.
Fillets Corner curves.
Finite element analysis (FEA) A way to test your material (element) of choice under forces on a computer model.
Footprint The area of land a building takes up.
Force A push or pull on an object causing it to change velocity (to accelerate) or change shape.
Fossil fuels Non-renewable resources, including coal, oil and natural gas, that can be burned to release energy.
Gear train Two or more gears arranged in series.
GRP Glass-reinforced plastic, also called fibreglass. A mix of fibreglass and epoxy resin.
Hatching A series of 45° parallel lines that are separated by an appropriate distance (e.g. 4 mm) to show where a solid object has been cut.
HSS High-speed steel. A type of steel that is very hard and contains a higher percentage of carbon than standard high-carbon steels. It has added elements like tungsten, molybdenum and vanadium. It is often used for cutting tools such as drill bits and saw blades. While HSS is sometimes referred to as a high-carbon steel, due to its carbon content, it is a specific and distinct type of steel known for its excellent cutting and tool performance, especially at high speeds.
Hydrochromic A substance that changes colour according to the level of water; *hydro* = relating to water, *chromic* = relating to colour.
Innovation From the Latin *innovare*, meaning to make new. To take something that already exists and improve it.
Interrelate To communicate with other people or, when talking about component parts, to work together.
Jig A workpiece-holding device that enables an operation such as drilling to be carried out repeatedly and safely.
Joules (J) The unit used to measure energy or work.

Key features Relevant pieces of information.
LED Light-emitting diode; an electrical component that gives off or emits light.
Leyden jar Also known as a Leiden jar, this is a simple glass jar with a foil-lined interior and exterior, used to store electric charge, much like a battery.
Linisher A machine that improves the flatness of a surface by sanding or polishing it. It is also known as a linish grinder.
Load The physical stress on a mechanical system or component.
Malleable Pliable, easy to shape without cracking or breaking.
Mandrel The long, cylindrical part of a rivet.
MDF MDF, or medium-density fibreboard, is a man-made board used for many applications such as furniture and internal building projects.
Measurement The size, length or amount of something, as established by measuring.
Measuring To discover the exact size, amount, etc., of something, or to be of a particular size.
Molten When a solid is heated to the point at which it becomes a liquid, it is said to be molten. Derives from the old English 'meltian', which means 'become liquid'.
Monomers Atoms or small simple molecules. The term comes from the Greek words *mono* = one and *meros* = part.
Non-destructive testing (NDT) A non-intrusive way of testing materials.
Non-renewable When describing a source of energy, it means that the source is finite and will eventually run out.
Obsolescence A product is created to become obsolete – old-fashioned or out of date.
Optimal The absolute best.
Orthographic projection In engineering, this is a means of representing different views of an object by projecting it onto a plane or surface.
Oxidisation In the process of oxidisation, steel/iron surfaces react with oxygen from the atmosphere and create ferric oxides (for example, rust).
Parallel Two things are said to be parallel if they are side by side and have the same distance between them continuously. If two lines are parallel, they will never meet.
Pattern A 3D copy of the item that is going to be cast. The sand is packed around the pattern and then the pattern is removed, creating a cavity that will be filled with the molten metal.
PCB A printed circuit board. This type of circuit board is printed using computer-aided manufacture.
Perpendicular Something that is at 90° to a given line.
Perspective Your point of view when looking at an object.
Photochromic A substance that changes colour according to the light; *photo* = relating to light, *chromic* = relating to colour.
Photons Light particles.
Pigment A coloured molecule.
Plan view A view from above, also known as a bird's-eye view.
Planes The X, Y and Z axes (directions) in which you create.
Planish To flatten, smooth or polish (metal) by rolling or hammering it.
Planishing A metalworking technique that involves finishing the surface of sheet metal by finely shaping and smoothing it.
Platen The part of the vacuum-forming machine that acts as a shelf. It has holes and can be raised or lowered as necessary.
Polymerisation The industrial process used to create plastics from naphtha. In scientific terms, polymerisation is the name of the process in which small molecules – called monomers – are joined together chemically to make a long, chain-like molecule called a polymer. Plastics are also called polymers.
Polymers Large molecules made up of many monomers joined together. Polymer is the scientific name for a plastic. *Poly* is the Greek word for many.

Glossary of key terms

Prism A solid shape that has two ends of the same shape and size. The length of a prism can vary.

Product analysis Looking at, feeling and maybe using a product to see how it works.

PVC A type of thermoplastic. Its full name is polyvinyl chloride.

Ratio The relationship between two groups or amounts that expresses how much bigger one is than the other.

Renewable energies Sources of energy that can be renewed naturally such as wind, solar, geothermal and tidal.

Representations Views.

Research The process of finding things out.

Resistance A material is said to be resistant if it resists a change.

Resistor An electrical component that can be used in a circuit to reduce/slow down the current in it.

Risk matrix A diagram used to define the level of risk by considering the category of probability or likelihood of an incident occurring against the category of consequence severity.

RPM Revolutions per minute. This is a measure of how fast a machine spins.

Runners A channel that guides the molten material from the sprue to the individual parts or cavities within the mould. It branches out from the sprue to distribute the material. The runner system helps evenly distribute the molten material to different parts of the mould. It ensures that each part receives the required amount of material for proper formation. Like the sprue, the runner is also considered waste material in relation to the final product but they can be reused.

Scribe To mark out.

Semiconductor Materials can be divided into three categories related to their ability to conduct electricity:
- high conductivity = conductor (e.g. metals)
- intermediate conductivity = semiconductor (e.g. silicon)
- low conductivity = insulator (e.g. plastics).

Sequencing Putting tasks into the correct order.

Slag The waste material that is left when smelting or refining metals from their ores.

SMA Shape memory alloy. A metallic alloy with 'memory'.

Sprue A hollow channel through which molten material (metal or plastic) is poured into a mould. It is the primary feed channel for the molten material to enter the mould cavity. The sprue allows the molten material to flow from the source to the mould, filling the mould cavity and creating the desired product. In the context of the final product, the sprue is considered waste material because it is typically removed from the finished product. However, sprues can often be reused.

Stock-form The standardised size and shape in which a material is available from suppliers.

Strength The capacity of an object or substance to withstand great force or pressure.

Sustainability In general terms, the ability to maintain or sustain something at a particular level. In terms of engineering and product design, sustainability relates to the creation of products with a minimal impact on the environment.

Sustainable A product is sustainable if its manufacture can be maintained (for example, if it is made from renewable resources).

Swarf Small pieces of metal or debris produced as a result of the machining process.

SWOT The acronym SWOT stands for:
- strengths: identifies good points about your project/idea
- weaknesses: identifies areas of your project/idea that could make it fail
- opportunities: identifies areas of your project/idea that you could exploit
- threats: identifies areas that could cause trouble for the project/idea.

Tailstock The part of the centre lathe that sits towards the rear of the machine and holds various useful tools such as chucks, drill bits and centre guides. It can also support longer workpieces to stop them from wobbling when they are being rotated.

Tap wrench A device for holding a tap. A tap wrench has arms with a textured surface to improve grip and can be rotated by hand.

Target market The group of users you will be designing for.

Technical drawings The common term used for third-angle orthographic projections.

Tension A force that occurs when something is being pulled or stretched.

Thermal Relating to or caused by heat.

Thermochromic A substance that changes colour according to the temperature; *thermo* = relating to temperature, *chromic* = relating to colour.

Tolerance An allowable amount of variation of a specified quantity, especially in the dimensions of a machine or part.

Top-down construction When constructing a building with basement floors you can complete the structure of the higher floors before excavating and constructing the lower floors.

TPI Teeth per inch. The number of teeth a saw blade has per inch of length.

Trammel A trammel of Archimedes (also known as an ellipsograph) is a device that can be used to draw ellipses. The trammel method can also be replicated by using a piece of paper and a major and minor axis of the ellipse to be drawn.

UPVC Unplasticised polyvinyl chloride. This hard form of PVC is often used for doors and windows.

Urea formaldehyde polymer A hard, slightly brittle plastic used for electrical casing/housing.

Vanishing points Lines that disappear into the distance.

Velocity The speed of something in a given direction.

Views Views are also known as elevations.

Weighted lines These lines define the object you are drawing, making it easier to see which lines to keep and which to erase. They are drawn with a soft (B) pencil or a fine-liner.

Photo credits

p. 2T © Monkey Business/stock.adobe.com; p. 2BL © ISO – International Organization for Standardization; p. 3 © The British Standards Institution 2024; p. 27BR © stockphoto-graf/Adobe Stock; p. 27BL © Pako/stock.adobe.com; p. 27BC © Maksym Yemelyanov/stock.adobe.com; p. 28T © frog/stock.adobe.com; p. 28B © raeva/stock.adobe.com; p. 29T © Marcus Harrison – textures / Alamy Stock Photo; p. 29B © hodim/stock.adobe.com; p. 41T © Yay Images/stock.adobe.com; p. 41CTR © photomelon/stock.adobe.com; p. 41C © Freedom Life/stock.adobe.com; p. 41CBR © Stephen/stock.adobe.com; p. 41BR © JLindsay/stock.adobe.com; p. 42TL, 52B(4) © terex/stock.adobe.com; p. 42CTL © Art of Success/stock.adobe.com; p. 42R © cristianstorto/stock.adobe.com; p. 42CBL © Christopher Dodge/stock.adobe.com; p. 42BL © krasyuk/stock.adobe.com; p. 43TL © fefufoto/stock.adobe.com; p. 43TC © Kj/stock.adobe.com; p. 43TR © Edward Westmacott/stock.adobe.com; p. 43BR © Artem Ogurtsov/Shutterstock.com; p. 43BL © Aleksandr Ugorenkov/stock.adobe.com; p. 44BL © Rinku Dua/Shutterstock.com; p. 44TL, 52B(3) © tclaxton/stock.adobe.com; p. 44CL © PHILL THORNTON PHOTO/stock.adobe.com; p. 44C © Cliff/stock.adobe.com; p. 44CR © Stefan/stock.adobe.com; p. 44BR © R.Moore/Shutterstock.com; p. 45TR © Vladimir Vydrin/stock.adobe.com; p. 45CTR, 52B(2) © Sergey Ryzhov/stock.adobe.com; p. 45CR © trongnguyen/stock.adobe.com; p. 45CBR © Piotr/stock.adobe.com; p. 45BL © russell witherington/stock.adobe.com; p. 45BC © kenary820/stock.adobe.com; p. 45BR © Cliff Day/Shutterstock.com; p. 46T © Cliff Day/Shutterstock.com; p. 46C, 52B(5) © vj/stock.adobe.com; p. 46B © tclaxton/stock.adobe.com; p. 47TR © Stocksnapper/stock.adobe.com; p. 47BR © troy/Shutterstock.com; p. 47L © Kunertus/Shutterstock.com; p. 47C © michaklootwijk/stock.adobe.com; p. 48TL © Mikhail Nesytykh/stock.adobe.com; p. 48BR © mehizm/stock.adobe.com; p. 49T © Veaceslav/stock.adobe.com; p. 49C © bizoo_n/stock.adobe.com; p. 49B © Winai Tepsuttinun/stock.adobe.com; p. 50(1) © smuki/stock.adobe.com; p. 50(2) © Andrzej Tokarski/stock.adobe.com; p. 50(3) © Oleksandr Dorokhov/stock.adobe.com; p. 50(4), 52B(6) © SERGEY/stock.adobe.com; p. 50(5), 52B(1) © Charles Risen/stock.adobe.com; p. 50(6) © Yaroslav/stock.adobe.com; p. 51TR © ded/stock.adobe.com; p. 51CTL © Rawf8/stock.adobe.com; p. 51CTR © VladKK/Shutterstock.com; p. 51CBL, 52B(7) © Evgenii Kurbanov/Shutterstock.com; p. 51CBR © VladKK/Shutterstock.com; p. 51TL © apithana/stock.adobe.com; p. 51CTL © kaw/stock.adobe.com; p. 51CBL © Tetiana Romaniuk/stock.adobe.com; p. 51BL © vladvm50/stock.adobe.com; p. 52R © pawelg601/stock.adobe.com; p. 53L © Andrey Eremin/Shutterstock.com; p. 53TR © Anselm Kempf/Shutterstock.com; p. 53CR © JT Jeeraphun/stock.adobe.com; p. 53BR © Anselm Kempf/Shutterstock.com; p. 54TL © Jordanhill School D&T Dept/Flickr/CC BY 2.0 DEED; p. 54CL © Steven/stock.adobe.com; p. 54C © mailsonpignata/stock.adobe.com; p. 54CR © Pixel_B/stock.adobe.com; p. 54BL © Pixel_B/stock.adobe.com; p. 54BR © KPixMining / Alamy Stock Photo; p. 55R, 62(3) © David J. Green / Alamy Stock Photo; p. 55C © Dmitry Kalinovsky/Shutterstock.com; p. 55L © Armen Khachatryan/Shutterstock.com; p. 56, 62(4) © Vereshchagin Dmitry/Shutterstock.com; p. 57T © patrikslezak/stock.adobe.com; p. 57B © Eimantas Buzas/stock.adobe.com; p. 58TL, 62(2) © Aleksandr Matveev/stock.adobe.com; p. 58CL © Chawran/Shutterstock.com; p. 58BL © Mischoko/stock.adobe.com; p. 58CL © Springfield Gallery/stock.adobe.com; p. 58CR © axpitel/stock.adobe.com; p. 58R © danishch/stock.adobe.com; p. 59TR, 62(1) © Creative Commons Attribution 2.0 Generic via Wikimedia Commons; p. 59CT © mofaez/Shutterstock.com; p. 59CBL © Dmitry Vereshchagin/stock.adobe.com; p. 59CBR © Gresei/stock.adobe.com; p. 59B © Aleksandr Matveev/stock.adobe.com; p. 60T © very_ulissa/stock.adobe.com; p. 60CL © François Mathieu/stock.adobe.com; p. 60CR © romankrykh/stock.adobe.com; p. 60B © Kadmy/stock.adobe.com; p. 61TL © Picunique/stock.adobe.com; p. 61TR, 62(6) © steve/stock.adobe.com; p. 61TR © stockphoto-graf/stock.adobe.com; p. 61CL © Carlos André Santos/stock.adobe.com; p. 61CR © natalieina17/stock.adobe.com; p. 61BL © Christopher Dodge/stock.adobe.com; p. 61B, 62(5) © Rony Zmiri/stock.adobe.com; p. 62BR © Rutmer/stock.adobe.com; p. 63T © DC Studio/stock.adobe.com; p. 63B © Summit Art Creations/stock.adobe.com; p. 64T, 66TL © R_boe/stock.adobe.com; p. 63C, 66TCL © pressmaster/stock.adobe.com; p. 63B © tiero/stock.adobe.com; p. 65TL © Olga/stock.adobe.com; p. 65TR, 66TR © Александр Ивасенко/stock.adobe.com; p. 65C, 66TCL © Sergey Ryzhov/stock.adobe.com; p. 65CL © Vladimir_Ya/stock.adobe.com; p. 65CR © oyoo/stock.adobe.com; p. 66CL © zadiraka vladislav/stock.adobe.com; p. 65BL © Dmitry Zimin/stock.adobe.com; p. 66(1) © sasimoto/stock.adobe.com; p. 66(2) © OLEKSANDR/stock.adobe.com; p. 66(3) © kozini/stock.adobe.com; p. 66(4) © philip kinsey/stock.adobe.com; p. 66(5) © pakphoto/stock.adobe.com; p. 72 © kerkezz/stock.adobe.com; p. 73 © Sebastian Duda/stock.adobe.com; p. 75T © Aleksandr Kondratov/stock.adobe.com; p. 75B © mkfilm/Shutterstock.com; p. 76T © Creative Commons CC0 1.0 Universal Public Domain Dedication via Wikimedia Commons; p. 76C © Jeniffer Fontan/Shutterstock.com; p. 76B © Rinku Dua/Shutterstock.com; p. 78T © ALAN OLIVER / Alamy Stock Photo; p. 78C © Jim West / Alamy Stock Photo; p. 78B © nordroden/stock.adobe.com; p. 80T © philip kinsey/stock.adobe.com; p. 80C © Elegant Solution/Shutterstock.com; p. 80B © rukawajung/stock.adobe.com; p. 81T © alexey_arz/stock.adobe.com; p. 81C © sir270/stock.adobe.com; p. surasak/stock.adobe.com; p. © Igor/stock.adobe.com; p. 81B © grigvovan/stock.adobe.com; p. 84TL © mkos83/stock.adobe.com; p. 84CTL, 85R © Acik/stock.adobe.com; p. 84CBL © doomu/stock.adobe.com; p. 84BL, 85L © kitthanes/stock.adobe.com; p. 86TL, 88(3) © eNJoy Istyle/stock.adobe.com; p. 86TC, 88(6) © Raymond Orton/stock.adobe.com; p. 86TR, 88(1) © Liza/stock.adobe.com; p. 86B, 88(4) © Pako/stock.adobe.com; p. 86CL, 88(8) © OZMedia/stock.adobe.com; p. 86BL, 88(5) © OZMedia/stock.adobe.com; p. 87B, 88(2) © Александр Довянский/stock.adobe.com; p. 87TL © Arpad Nagy-Bagoly/stock.adobe.com; p. 101T © Maksim Bukovski/stock.adobe.com; p. 101CT © cristianstorto/stock.adobe.com; p. 101CB © 3drenderings/stock.adobe.com; p. 101B © hsagencia/stock.adobe.com; p. 102(1) © Reidos/stock.adobe.com; p. 102(2) © Vladislav12/stock.adobe.com; p. 102(3) © theerakit/stock.adobe.com; p. 102(4) © WINDCOLORS/stock.adobe.com; p. 102(5) © Veniamin Kraskov/stock.adobe.com; p. 102(6) © Denis Dryashkin/stock.adobe.com; p. 102(7) © KPixMining/Shutterstock.com; p. 102(8) © Krasowit/Shutterstock.com; p. 103(1) © photology1971/stock.adobe.com; p. 103(2) © Timothy Hodgkinson/stock.adobe.com; p. 103(3) © Denis Rozhnovsky/stock.adobe.com; p. 103(4) © BM Digital/stock.adobe.com; p. 103(5) © robertkoczera/stock.adobe.com; p. 103(6) © bm_photo/stock.adobe.com; p. 103(7) © ra3rn/stock.adobe.com; p. 104T © Igor/stock.adobe.com; p. 105B © ANATOL/Shutterstock.com; p. 105; p. 108 © Tj/stock.adobe.com; p. 112 © New Africa/stock.adobe.com; p. 114TR © SHEILA TERRY / SCIENCE PHOTO LIBRARY; p. 114BR © gerasimenuk/stock.adobe.com; p. 115 © Yomka/stock.adobe.com; p. 116 © Alex/stock.adobe.com; p. 117 © rawpixel.com/stock.adobe.com; p. 118 © Родион Бондаренко/stock.adobe.com; p. 128TL © Nanda/stock.adobe.com; p. 128TR © SimoneO/stock.adobe.com; p. 128B © piyaphunjun/stock.adobe.com; p. 129T © artisan263/stock.adobe.com; p. 129B © Andrzej Tokarski/stock.adobe.com; p. 134 © archerix/stock.adobe.com; p. 135(1) © nmcandre/stock.adobe.com; p. 135(2) © James King-Holmes / Alamy Stock Photo; p. 134(3) © nordroden/stock.adobe.com; p. 134(4) © Sergey Ryzhov/stock.adobe.com; p. 135(5) © Kadmy/stock.adobe.com; p. 141 © Stephen/stock.adobe.com; p. 142 © The Reading Room / Alamy Stock Photo; p. 143 © DMH/stock.adobe.com; p. 144 © Gorodenkoff/stock.adobe.com; p. ; p. 147T © Gorodenkoff/Shutterstock.com; p. 147B © Yohan Lafond/stock.adobe.com; p. 148T © JackF/stock.adobe.com; p. 148B © charles/stock.adobe.com; p. 149 © engel.ac/stock.adobe.com; p. 151 © Rob Byron/stock.adobe.com; p. 152 © Kev Gregory/Shutterstock.com; p. 153T © Scanrail/stock.adobe.com; p. 153B © DimaBerlin/stock.adobe.com; p. 154TL © Mierna/Shutterstock.com; p. 154C © 3drenderings/stock.adobe.com; p. 154BL © somemeans/stock.adobe.com; p. 156T © mike/stock.adobe.com; p. 156B © Soonthorn/stock.adobe.com; p. 157 © 수동 김/stock.adobe.com; p. 158 © klyaksun/stock.adobe.com; p. 161 © Gorodenkoff/stock.adobe.com; p. 162T © yuliachupina/stock.adobe.com; p. 162B © didiksaputra/stock.adobe.com; p. 164TL © monticelllo/stock.adobe.com; p. 164BR © Tina/stock.adobe.com; p. 169 © Alex/stock.adobe.com; p. 170(1) © Juli/stock.adobe.com; p. 170(2) © squeebcreative/stock.adobe.com; p. 170(3) © Luka/stock.adobe.com; p. 170(4) © Kitch Bain/stock.adobe.com; p. 170(5) © Maksim Bukovski/stock.adobe.com; p. 170(6) © OlegDoroshin/stock.adobe.com; p. 170(7) © Andrew Gardner/stock.adobe.com; p. 171 © Trucafort/stock.adobe.com; p. 173T © Paul/stock.adobe.com; p. 173CT, 174(4) © chas53/stock.adobe.com; p. 173CB, 174(5) © ELENA/stock.adobe.com; p. 173BR, 174(6) © pixelrobot/stock.adobe.com; p. 173L, 174(7) © Kuzmick/stock.adobe.com; p. 174(1) © Martina Simonazzi/stock.adobe.com; p. 174(2) © butus/stock.adobe.com; p. 174(3) © ELENA/stock.adobe.com; p. 174B © tapong117/stock.adobe.com; p. 175(T) © WavebreakmediaMicro/stock.adobe.com; p. 175(1) © Thomas N./stock.adobe.com; p. 175(2) © rukawajung/stock.adobe.com; p. 175(3) © BillionPhotos.com/stock.adobe.com; p. 175(4) © Наталья Добровольська/stock.adobe.com; p. 175(5) © ValentinValkov/stock.adobe.com; p. 175(6) © Nomad_Soul/stock.adobe.com; p. 176T © eyalg_115/stock.adobe.com; p. 176C © djekill2007/stock.adobe.com; p. 176B © Martina007/stock.adobe.com; p. 177T © Tania/stock.adobe.com; p. 177C © claudia veja images/Shutterstock; p. 177B © epitavi/stock.adobe.com; p. 178 © kontur-vid/stock.adobe.com; p. 179TR © marcoscastro/stock.adobe.com; p. 181T © Chrispo/stock.adobe.com; p. 181C © vvoe/stock.adobe.com; p. 181B © nordroden/stock.adobe.com; p. 182TL © tarasov_vl/stock.adobe.com; p. 182L © Alex/stock.adobe.com; p. 182CL © Aleksandrkozak/Shutterstock.com; p. 182CR © stockphoto-graf/stock.adobe.com; p. 182R © majeczka/stock.adobe.com; p. 183 © Funtay/stock.adobe.com; p. 184 © nuttawutnuy/stock.adobe.com; p. 192 © FERNANDO/stock.adobe.com.

Index

3D printing 64, 87, 144
3D shapes *see* isometric drawing; volume
6Rs of sustainability 161–2
ACCESS FM model 110–11, 142
acrylic 173
additive manufacturing 64
adhesives 80, 81, 174
aeroplanes 152
aesthetics (product design) 108
Allen keys 52
alloying agents/elements 170
alloys 76, 168–9, 177
aluminium 168, 169
ammeters 103
angle finders 42
angle grinders 59
angle plates 76
angles
   drawings 4, 6, 8, 15, 22, 23, 42–3
   marking-out 75
annotations 108, 142, 175–6
anodising 86
anthropometrics 109, 111
anvils 50
area (calculation) 119–21
aroma pigments 178
assembly tools 52
automobiles 150–1
axes in diagrams 4
axial loads 134
axonometric perspective 9
Bakelite (phenolic resin) 175
bar charts/graphs 137
batteries (cells) 102
bench drills 56, 68
bench grinders 59
billets 82
biodegradability 160, 163
bioplastic 161
blow moulding 84
blowtorches (oxy-acetylene) 77
blueing 86
boards (wood) 176
borders to drawings 20
boring (lathes) 54
bought-in-components 31
bradawls 44
brass 170
brazing 77–8
bridges 147–8
briefs (projects) *see* design briefs
British Standards Institution *see* BSI
brittleness 166
bronze 169

BSI standards 2–3, 4, 13
buffing/polishing machines 59
bullet-point design briefs 31, 106–7
cable-stayed bridges 148
CAD (computer-aided design)
   accessibility 39
   advantages/disadvantages 63
   design drawings 4, 15, 63, 141
   design models 63, 144
   in structural engineering 62
   testing processes 186
   use with CAM 63–4
callipers 41–2
CAM (computer-aided manufacturing) 39, 62, 63–4
capacitors 101
carbon fibre 176
carpenter's try squares 43
cars 150–1
cells (batteries) 102
centre finders 42
centre lathes 36, 52–4, 56, 58, 65, 71
centre punches 44, 75
chasing (threads) 51
chemicals, health and safety 72–3
chipboard 176
chisels 61
chucks 53, 54, 58
circles 9, 120–1
circuit diagrams 104, 154
clamps 45–6
client needs 109, 111
clothing (protection) 66, 67
CNC (computerised numerical control) 39, 64–5
combination squares 42, 75
communication of designs 142–3
components 101–5, 112–14
composite materials 175–6, 177
compound shapes, area 121
compression force 147–8, 149
compression moulding 28
compressive strength 166, 181
condensed design briefs 31, 106
conductivity 105, 166, 181
construction lines 4, 5–6, 7, 18
contingency planning 38–9
copper 168, 169, 170
corded/cordless drills 57
corrosion resistance 85, 166, 181, 186
COSHH regulations 72–3
costings 119, 160
crates (3D box) 5–6
cubes and cuboids 122

curves, isometric 9
customer needs 109, 111
cut plane lines 18, 19, 20
cutaway drawings 12, 151, 176
cutting lists 33–4
cutting tools
   CNC cutters 64, 65
   for metal 46, 51, 52, 53–4, 59–60
   for plastic 46, 47, 58, 64
   for wood 46–7, 60, 61
data sheets 35–6, 56–7, 71–2
datum points 123
deburring tools 50
design briefs 30–1, 105–8, 117, 139
design modification 143–4, 159, 162
design specifications 88–9, 115–17, 139
designing products
   see also project planning
   design briefs 31, 105–8, 117, 139
   exploded views 12
   features and components 101–5
   product research 108–14
   mathematical techniques 117–38
   specifications 88–9, 115–17, 139
   modification 143–4, 159, 162
   testing materials 183
   evaluation of project 138–40
   product lifecycle 157–64
destructive tests 183, 184–5
dies (casting) 82
dies (thread cutting) 51, 54
dimensions 16–17, 20, 23, 32, 143
diodes 101, 102
disc grinders 59
disposal of products 159–60
distribution of goods 159
dividers 43, 75
doming block and punch set 50, 81
DPIs (inspections) 89
drawings
   use of CAD 4, 15, 63, 141
   cutaway 12, 151, 176
   in development 141–2, 143, 144
   elevations 13–16, 22
   equipment 3, 4, 5, 22
   exploded views 12–13
   isometric drawings 4–13, 22, 141
   lines 3–7, 13, 18–20, 21–4
   orthographic projections 13–25, 31–3, 142
   sketches 10, 11, 141
   standards 2–3
drill bits 51, 58
drilling machines 44, 54, 56, 57, 58, 68
drills (hand drills) 57

driver/driven gears 133, 134, 136
driving/driven shafts (pulleys) 132
drop forging 82
ductility 166, 181, 185
durability 105
duralumin 170
dye penetrant testing 184
eddy current testing 184
elasticity 166, 182, 186
electrical components 101–4, 109, 160
electrical conductivity 166, 181, 185
electrical values 104
electrical motors 103
electrodes (in welding) 78, 79
electroluminescent materials 179
electronic components 101–4, 109, 160
electronic engineering
    achievements and examples 153–4
    components 101–4, 109
    Ohm's law 127
    recycling directives 160
electro-rheostatic (ER) fluids 179
elevations 13–16, 19–20, 22–4
ellipses in drawing 9–11
ellipsographs 10
enamelling 86
end milling 56
energy efficiency 126
engineering developments
    achievement examples 146–55
    environmental policies 156–60
    sustainable materials 160–4
engineering drawings *see* orthographic projections
engineering equipment *see* equipment
engineering processes
    marking-out 74–6
    joints and fixings 76–81
    shaping processes 49–51, 53, 54, 55, 61, 81–5
    finishing processes 85–7
    use of CAM 63–4
    environmental impact 159–60
engineering solutions
    developing ideas 140–2, 143–5
    communication of designs 142–3
    models for designs 144–5
    evaluation of ideas 139–40
engineer's blue 43, 75
engineer's try squares 42, 75
environmental degradation 166, 182
environmental issues
    product lifecycle 110, 157–60
    renewable energy 156–7
    sustainability 110, 157–64
epoxy resin 80, 174
equipment
    see also CAD; hand tools; machine tools
    assembling tools 52

contingencies 39
cutting tools 46–7, 51–2, 53–6, 58, 59–60, 61, 64–5
    for drawings 3, 5
drills and drilling 54, 56–8
finishing tools 59, 60
holding tools 44–6
marking tools 42–4
measuring tools 43, 44
PPE 66–7
project planning 34, 35
shaping tools 49–51, 53, 54, 55, 61
ergonomics 109, 111
estimation 118
etching (printed circuit boards) 62
evaluation of projects
    design solutions 138–40
    personal practice 90
    product outcome 88–9, 145
expendable moulds 83
exploded views 12–13
extension lines 18, 19
extrusions (3D shapes) 7
extrusions (materials) 27, 28
face milling 55
facing off 53
factor of safety (FoS) 183
FEA testing 144, 186
feed and speed rates 36, 56–7, 71
ferrous metals 167–8
filing tools 47–9
fillets (corner curves) 11
final inspections 89
finishing processes 49, 59, 60, 82, 83, 85–7
finite element analysis 144, 186
first angle method 14, 15
fixings 76–81, 104–5
flat marking 75
flux (soldering and welding) 77, 79
Ford Model T car 151
forging 82
Formica (urea-methanal resin) 175
FoS (factor of safety) 183
four-centre ellipses 10–11
fractional distillation 171–2
friction 105
fulcrum (levers) 128–9
fuses 102
galvanising 86
Gantt charts 36–7
gauge measurement 27, 82
gauges (marking) 43, 44
gear ratio 136
gear trains 134
gears 134–6
geothermal power 157
global company policies 162–3
glue 80
granules (plastics) 177

graphs 137–8
grinders 59
grooving 54
GRP, composite material 176
guillotines 60
hammer drills 57
hammers 49
hand drills 57
hand files 47–8
hand shears 50
hand tools
    assembly tools 52
    cutting 46–7, 51
    drawbacks 156
    drills 57
    filing 47–9
    holding 44–6
    marking 42–4
    measuring 41–2
    shaping and improving 49–51, 61
hardness 166, 181, 184
harmful substances regulations 72–3
hatching 12, 19, 20
hazards 68–9, 72–3
HC polymers 179
health and safety 65–73
    COSHH regulations 72–3
    data sheets 35–6, 71–2
    guidelines 73
    highlighting in job sheets 35
    PPE 66–7
    risk assessment 67–9
    safety signage 69–71
hex keys 52
high-impact polystyrene (HIPS) 173
histograms 137
holding tools 44–6
HSS (high-speed steel) 58
hydrochromic materials 178
hydropower 150, 157
idler gears 134
imperial system 118
Industrial Revolution 156
injection moulding 28, 83–4
internal combustion engine 151
International Organization for Standardization 2–3, 4, 13, 163
iPhones 154
iron 167, 168, 170
ISO standards 2–3, 4, 13, 163
isometric drawing 4–13, 141
    3D shape construction 5–7
    angles, circles, curves and ellipses 4, 6, 8–11
    axonometric perspective 9
    for client communication 141
    cutaway drawings and exploded views 12–13, 151, 176
    in orthographic projection 22
    projection explained 4, 141

Index

isometric grid paper 5
isometric projection 4, 141
   see also isometric drawing
jet propulsion 152
jigs 44, 46
job sheets 34–5
joints and fixings 76–81, 104–5
Kevlar 177
knock-down fittings 81
knurling 36, 54
laser cutters 64
lathes 36, 52–4, 56, 58, 61, 65
LDRs 103
lead 169
LEDs 102
levers 128–9
light-dependent resistors 103
light-emitting diodes 102
limitations of product 111
line graphs 137
linear motion 109
lines (drawing) 4, 5–7, 13, 18–20, 21–4
linishers 60–1
machine drills 56
machine tools
   see also CAM
   buffing/polishing 59
   CNC machines 64–5
   data sheets 35–6, 56–7, 71–2
   drills 56–8
   grinders 59
   guillotines 60
   lathes 53–4, 61, 65
   milling machines 55–6, 65
   routers 61, 65
   sanders 60–1
   saws 59, 60
magnetic particle inspections 184
magneto-rheostatic (MR) fluids 179
malleability 166, 181, 185
mallets 50
mandatory signs 70
manufacturing *see* engineering processes;
   planning manufacture
marking-out tools 42–4, 75–6
materials
   see also metal and metals; plastic and
      plastics; wood
   engineering properties 180–2
   physical properties 165–6
   stock forms 26–9, 33, 35, 39
   sustainability 160–1
   testing for product design 183–6
   types and uses 167–79
   use of resources 159
mathematical techniques 117–38
   areas 119–21
   for electronics 127
   energy efficiency 126

financial costings 119
   graphs 137–8
   mean (average) 125
   measuring 117–19
   scale 123–4
   volumes 122–3
MDF (fibreboard) 176
mean (average) 125
measurable criteria 116
measuring 41–2, 43, 44, 117–19
mechanical advantage 128–36
mechanical engineering 109, 150–2
mechanisms 109
melamine formaldehyde 174
metal and metals
   alloying agents 170
   centre lathe processes 52–4
   cutting tools 46, 51–2, 53–4, 59–60
   finishing processes 59, 82, 86
   joints and fixings 76–9, 80
   marking-out 42, 43, 44, 75–6
   milling processes 55–6
   properties 167–70
   recycling 164
   rpm data sheets 36, 56–7, 71
   shaping 49–51, 81–3
   smoothing tools 47–9
   stock forms 27
   uses 168, 169, 170
metal arc welding 79
metal inert gas welding 78
metric system 118
micrometers 41
MIG welding 78
Millau Viaduct 148
milling machines 55–6, 57, 65, 71
mitre squares 43
models for designs 144–5
mouldings (timber) 28
moulds
   metal casting 82–3
   for plastic 28, 83–5
multimeters 42
nanomaterials 179
nitinol 177
'no danger' signs 70, 71
non-destructive tests 184
non-ferrous metals 168–70
nuts and bolts 51–2, 81, 104
nylon 173
obsolescence 140, 162
odd-leg callipers 41–2, 75
offset section lines 20
Ohm's law 127
one plane drawings 8, 11
optimal solutions 140
orthographic projections 13–25
   components 20–1
      dimensions and scale 16–17, 20, 21

for finished designs 142
   first- and third-angle 14–16, 22–3
   interpreting in projects 31–3
   lines 18–20, 21–4
   success criteria 25
   symbols 14–15, 16–17
   views (elevations) 13–16, 22
oscillating motion 109
oxy-acetylene gas welding 78, 79
paint finishes 86
parallelogram area 120
parallels (bars) 76
parisons 84
parting 36, 53
parts lists 14, 21, 23, 32
patterns (casting) 83
PCBs 62, 88–90, 101
peripheral milling 55
permanent joints and fixings 76–9
personal protective equipment 66–7
perspective in drawing 9
phenol formaldehyde 175
photochromic materials 178
photovoltaic cells/surfaces 156, 161
physical models 145
pie charts 138
piezoelectric materials 178
pillar drills 56, 57, 68, 71, 72
planers 61
planes in 2D and 3D drawing 4, 7
planishing 49, 81
planning manufacture
   materials 26–9, 39
   time management 36–9
   use of CAD 63
plasma cutters 65
plastic and plastics
   adhesive bonding 80
   biodegradability 160, 163
   cutting tools 46, 47, 58, 64
   finishing 49, 59, 60, 83, 87
   marking-out 75
   production 171–2
   properties 171
   recycling 163, 164
   self-finishing 83
   shaping and moulding 55, 83–5
   stock forms 28, 177
   types and uses 161, 172–5, 177–80
plastic dip coating 86
platens 85
pliers 50
polishing machines 59
pollution-absorbing bricks 161
polyester resin 175
polyesters 173
polymers/polymerisation 171–2
polymorph 177
polypropene 173

polystyrene (polyphenylethene) 173
polythene (polyethene) 173
pop riveting 80
powder coating 86
PPE 66–7
PPI (pre-production inspections) 89
printed circuit boards 62, 101
prioritised bullet-point briefs 31, 107
prisms, volume 123
product analysis 110–11
product design *see* designing products
product lifecycle 157–60
production levels 111
prohibition signs 70, 71
project planning
   see also designing products
   design briefs 30–1, 105–8, 117, 139
   interpreting drawings 31–3
   cutting lists 33–4
   job sheets and sequencing 34–5
   time management 36–7
   contingency planning 38–9
projection lines
   isometric 4, 5–7, 9, 13, 141
   orthographic 18–20, 21–4
prototypes 145
pulleys 129–32
pulsejet engines 152
PVC 173
QTC (composite) 177
qualitative data 116
quality/quality control 35, 88–9, 105
quality inspections 89
quantitative data 116
radial reaction loads 134
radiography testing 184
radios 154
rasps 49
ratchets and sockets 52
reciprocating motion 109
rectangle area 120
recycling/reuse 160, 162, 163–4
reinforced concrete 176
reliability of product 140
renewable energy 156–7
research in product design 108–14
resistors 101, 102, 103
respiratory protection (RPE) 67
reverse engineering 111–12
risk assessments 35, 67–9
risk matrix 68–9
rivets and riveters 80
rotary motion 109
rotational moulding 84
routers 61, 65
RPE (respiratory protection) 67
sacrificial metal 77
safety
   personnel *see* health and safety

product design 109, 183
sand casting 83
sanders 60–1
saws
   metalwork 59, 60
   woodworking 46–7
scale of drawings 21, 123–4
SCAMPER technique 143–4
screwdrivers 52
screws 81, 105
scribers 43, 75
scribing blocks 76
section lines 18, 19, 20
sections and extrusions 27, 28
self-evaluation of projects 90, 138–40
self-finishing plastics 83
self-healing material 161
semi-conductors 153
shaping
   processes 81–5
tools 49–51, 53–4, 55–6, 61
Shard (skyscraper) 149
sheet-metal benders 51
sheets (materials) 27
side grinders 59
signage (safety) 69–71
sketches 10, 11, 141, 143
sliding T-bevels 42
slot milling 55
smart factories 161
smart materials 177–9
solar power 156–7
soldering 76–7
spanners 52
speakers 103
Spitfire (aeroplane) 152
stain finishes 87
stainless steel 167, 168, 170
standards for drawings 2–3, 13
steel 168, 169
steel rules 41, 75
stick welding 79
stock forms 26–9, 33, 35, 39
structural engineering 146–9
success criteria 25, 88, 116, 139, 143
surface tables/plates 43, 76
sustainability 110, 157–64
sustainable concrete 160
switches 102
SWOT analysis 90, 145
symbols 14–17, 103, 163
taper turning 54
tapered reamers 50
taps and dies 51–2
technical drawings *see* orthographic
   projections
Teflon (polytetrofluoroethene) 173
temporary joints/fixings 80–1
tensile strength 166, 181, 185

tension force 147–8, 149
thermal conductivity 166, 181, 185
thermistors 102
thermochromic materials 178
thermoplastics 172–3
thermosetting plastics 174–5
third angle projections 14, 15–16, 22–3, 141
thread cutting 51, 54
three plane drawings 4, 5, 11, 13
tidal power 157
timber *see* wood
time management 36–7
title blocks 21, 22
tolerance 20, 33, 89
tools *see* equipment
top-down construction 149
touch-screen technology 154
toughness 166, 181, 185
transistors 101, 153–4
try squares 42–3
turning 53
two plane drawings 4, 5, 8
ultrasonic flaw detection 184
UPVC 173
urea formaldehyde 174
urea-methanal resin 175
user needs 109, 111
vacuum forming 84, 85
variable resistors 102
varnishing 87
V-blocks 44, 76
velocity ratio
   gears 136
   pulleys 131–2
vertical milling machines 55–6
viaducts 148
vices 45
views (elevations) 13–16, 19–20, 22–4
visual inspections 184
voltmeters 103
volume (measurement) 122–3
warning signs 70, 71
water wheels 150
water-jet cutters 65
WEEE recycling directive 160
weighted lines 4, 5, 6, 7, 13, 18
welding 67, 78–9
wet and dry paper 59
wind power 156
wood
   composite materials 176
   finishing processes 60, 86–7
   FSC certification 163
   joints and fixings 80, 81
   marking-out 75
   stock forms 28–9
   tools 45–7, 49, 60, 61
wood-turning lathes 61
wrenches 51, 52